www.EffortlessMath.com

... So Much More Online!

- ✓ FREE Math lessons
- ✓ More Math learning books!
- ✓ Mathematics Worksheets
- ✓ Online Math Tutors

SSAT Upper Level Math Prep 2020-2021

The Most Comprehensive Review and Ultimate Guide to the SSAT Upper Level Math Test

By

Reza Nazari & Ava Ross

All inquiries should be addressed to:

info@effortlessMath.com

www.EffortlessMath.com

ISBN: 978-1-64612-212-7

Published by: Effortless Math Education

www.EffortlessMath.com

Description

SSAT Upper Level Math Prep 2020 – 2021, which reflects the 2020 - 2021 test guidelines, is dedicated to preparing test takers to ace the SSAT Upper Level Math Test. This comprehensive SSAT Upper Level Math Prep book with hundreds of examples, abundant sample SSAT Upper Level mathematics questions, and two full-length and realistic SSAT Upper Level Math tests is all you will ever need to fully prepare for the SSAT Upper Level Math. It will help you learn everything you need to ace the math section of the SSAT Upper Level test.

Effortless Math unique study program provides you with an in-depth focus on the math portion of the exam, helping you master the math skills that students find the most troublesome. This SSAT Upper Level Math preparation book contains most common sample questions that are most likely to appear in the mathematics section of the SSAT Upper Level.

Inside the pages of this comprehensive SSAT Upper Level Math book, students can learn basic math operations in a structured manner with a complete study program to help them understand essential math skills. It also has many exciting features, including:

- ✓ Content 100% aligned with the 2020 SSAT Upper Level test
- ✓ Written by SSAT Upper Level Math instructors and test experts
- ✓ Complete coverage of all SSAT Upper Level Math concepts and topics which you will be tested
- ✓ Over 2,500 additional SSAT Upper Level math practice questions in both multiple-choice and grid-in formats with answers grouped by topic, so you can focus on your weak areas
- ✓ Abundant Math skill building exercises to help test-takers approach different question types that might be unfamiliar to them
- ✓ Exercises on different SSAT Upper Level Math topics such as integers, percent, equations, polynomials, exponents and radicals
- ✓ 2 full-length practice tests (featuring new question types) with detailed answers

SSAT Upper Level Math Prep 2020 – 2021 is an incredibly useful resource for those who want to review all topics being covered on the SSAT Upper Level test. It efficiently and effectively reinforces learning outcomes through engaging questions and repeated practice, helping you to quickly master Math skills.

About the Author

Reza Nazari is the author of more than 100 Math learning books including:
– **Math and Critical Thinking Challenges:** For the Middle and High School Student
– **ACT Math in 30 Days**
– **ASVAB Math Workbook 2018 - 2019**
– **Effortless Math Education Workbooks**
– **and many more Mathematics books ...**

Reza is also an experienced Math instructor and a test–prep expert who has been tutoring students since 2008. Reza is the founder of Effortless Math Education, a tutoring company that has helped many students raise their standardized test scores—and attend the colleges of their dreams. Reza provides an individualized custom learning plan and the personalized attention that makes a difference in how students view math.

You can contact Reza via email at:
reza@EffortlessMath.com

Find Reza's professional profile at:
goo.gl/zoC9rJ

Contents

Chapter 1:

Fractions and Mixed Numbers

Math Topics that you'll learn in this Chapter:

- ✓ Simplifying Fractions
- ✓ Adding and Subtracting Fractions
- ✓ Multiplying and Dividing Fractions
- ✓ Adding Mixed Numbers
- ✓ Subtracting Mixed Numbers
- ✓ Multiplying Mixed Numbers
- ✓ Dividing Mixed Numbers

Simplifying Fractions

- A fraction contains two numbers separated by a bar in between them. The bottom number, called the denominator, is the total number of equally divided portions in one whole. The top number, called the numerator, is how many portions you have. And the bar represents the operation of division.
- Simplifying a fraction means reducing it to lowest terms. To simplify a fraction, evenly divide both the top and bottom of the fraction by $2, 3, 5, 7, \ldots$ etc.
- Continue until you can't go any further.

Examples:

1) Simplify $\frac{12}{30}$

Solution: To simplify $\frac{12}{30}$, find a number that both 12 and 30 are divisible by.

Both are divisible by 6. Then: $\frac{12}{30} = \frac{12 \div 6}{30 \div 6} = \frac{2}{5}$

2) Simplify $\frac{64}{80}$

Solution: To simplify $\frac{64}{80}$, find a number that both 64 and 80 are divisible by.

Both are divisible by 8 and 16. Then: $\frac{64}{80} = \frac{64 \div 8}{80 \div 8} = \frac{8}{10}$, 8 and 10 are divisible by 2,

then: $\frac{8}{10} = \frac{4}{5}$ or $\frac{64}{80} = \frac{64 \div 16}{80 \div 16} = \frac{4}{5}$

3) Simplify $\frac{20}{60}$

Solution: To simplify $\frac{20}{60}$, find a number that both 20 and 60 are divisible by. Both are divisible by 20.

Then: $\frac{20}{60} = \frac{20 \div 20}{60 \div 20} = \frac{1}{3}$

Adding and Subtracting Fractions

- For "like" fractions (fractions with the same denominator), add or subtract the numerators (top numbers) and write the answer over the common denominator (bottom numbers).
- Adding and Subtracting fractions with the same denominator:

$$\frac{a}{b}+\frac{c}{b}=\frac{a+c}{b} \qquad \frac{a}{b}-\frac{c}{b}=\frac{a-c}{b}$$

- Find equivalent fractions with the same denominator before you can add or subtract fractions with different denominators.
- Adding and Subtracting fractions with different denominators:

$$\frac{a}{b}+\frac{c}{d}=\frac{ad+bc}{bd} \qquad \frac{a}{b}-\frac{c}{d}=\frac{ad-bc}{bd}$$

Examples:

1) Find the sum. $\frac{3}{4}+\frac{1}{3}=$

Solution: These two fractions are "unlike" fractions. (they have different denominators). Use this formula: $\frac{a}{b}+\frac{c}{d}=\frac{ad+cb}{bd}$

Then: $\frac{3}{4}+\frac{1}{3}=\frac{(3)(3)+(4)(1)}{4\times 3}=\frac{9+4}{12}=\frac{13}{12}$

2) Find the difference. $\frac{4}{5}-\frac{3}{7}=$

Solution: For "unlike" fractions, find equivalent fractions with the same denominator before you can add or subtract fractions with different denominators. Use this formula:

$\frac{a}{b}-\frac{c}{d}=\frac{ad-bc}{bd}$

$\frac{4}{5}-\frac{3}{7}=\frac{(4)(7)-(3)(5)}{5\times 7}=\frac{28-15}{35}=\frac{13}{35}$

Multiplying and Dividing Fractions

- Multiplying fractions: multiply the top numbers and multiply the bottom numbers.

  Simplify if necessary. $\frac{a}{b} \times \frac{c}{d} = \frac{a \times c}{b \times d}$

- Dividing fractions: Keep, Change, Flip

  Keep first fraction, change division sign to multiplication, and flip the numerator and denominator of the second fraction. Then, solve! $\frac{a}{b} \div \frac{c}{d} = \frac{a}{b} \times \frac{d}{c} = \frac{a \times d}{b \times c}$

Examples:

1) Multiply. $\frac{5}{8} \times \frac{2}{3} =$

Solution: Multiply the top numbers and multiply the bottom numbers.
$\frac{5}{8} \times \frac{2}{3} = \frac{5 \times 2}{8 \times 3} = \frac{10}{24}$, simplify: $\frac{10}{24} = \frac{10 \div 2}{24 \div 2} = \frac{5}{12}$

2) Solve. $\frac{1}{3} \div \frac{4}{7} =$

Solution: Keep first fraction, change division sign to multiplication, and flip the numerator and denominator of the second fraction.
Then: $\frac{1}{3} \div \frac{4}{7} = \frac{1}{3} \times \frac{7}{4} = \frac{1 \times 7}{3 \times 4} = \frac{7}{12}$

3) Calculate. $\frac{3}{5} \times \frac{2}{3} =$

Solution: Multiply the top numbers and multiply the bottom numbers.
$\frac{3}{5} \times \frac{2}{3} = \frac{3 \times 2}{5 \times 3} = \frac{6}{15}$, simplify: $\frac{6}{15} = \frac{6 \div 3}{15 \div 3} = \frac{2}{5}$

4) Solve. $\frac{1}{4} \div \frac{5}{6} =$

Solution: Keep first fraction, change division sign to multiplication, and flip the numerator and denominator of the second fraction.
Then: $\frac{1}{4} \div \frac{5}{6} = \frac{1}{4} \times \frac{6}{5} = \frac{1 \times 6}{4 \times 5} = \frac{6}{20}$, simplify: $\frac{6}{20} = \frac{6 \div 2}{20 \div 2} = \frac{3}{10}$

Adding Mixed Numbers

Use following steps for adding mixed numbers:

- ✓ Add whole numbers of the mixed numbers.
- ✓ Add the fractions of the mixed numbers.
- ✓ Find the Least Common Denominator (LCD) if necessary.
- ✓ Add whole numbers and fractions.
- ✓ Write your answer in lowest terms.

Examples:

1) Add mixed numbers. $3\frac{1}{3} + 1\frac{4}{5} =$

Solution: Let's rewriting our equation with parts separated, $3\frac{1}{3} + 1\frac{4}{5} = 3 + \frac{1}{3} + 1 + \frac{4}{5}$. Now, add whole number parts: $3 + 1 = 4$

Add the fraction parts $\frac{1}{3} + \frac{4}{5}$. Rewrite to solve with the equivalent fractions. $\frac{1}{3} + \frac{4}{5} = \frac{5}{15} + \frac{12}{15} = \frac{17}{15}$. The answer is an improper fraction (numerator is bigger than denominator). Convert the improper fraction into a mixed number: $\frac{17}{15} = 1\frac{2}{15}$. Now, combine the whole and fraction parts: $4 + 1\frac{2}{15} = 5\frac{2}{15}$

2) Find the sum. $1\frac{2}{5} + 2\frac{1}{2} =$

Solution: Rewriting our equation with parts separated, $1 + \frac{2}{5} + 2 + \frac{1}{2}$. Add the whole number parts:

$1 + 2 = 3$. Add the fraction parts: $\frac{2}{5} + \frac{1}{2} = \frac{4}{10} + \frac{5}{10} = \frac{9}{10}$

Now, combine the whole and fraction parts: $3 + \frac{9}{10} = 3\frac{9}{10}$

Subtract Mixed Numbers

Use the following steps for subtracting mixed numbers.

- ✓ Convert mixed numbers into improper fractions. $a\frac{c}{b}=\frac{ab+c}{b}$
- ✓ Find equivalent fractions with the same denominator for unlike fractions. (fractions with different denominators)
- ✓ Subtract the second fraction from the first one. $\frac{a}{b}-\frac{c}{d}=\frac{ad-bc}{bd}$
- ✓ Write your answer in lowest terms.
- ✓ If the answer is an improper fraction, convert it into a mixed number.

Examples:

1) Subtract. $3\frac{4}{5}-1\frac{3}{4}=$

Solution: Convert mixed numbers into fractions: $3\frac{4}{5}=\frac{3\times5+4}{5}=\frac{19}{5}$ and $1\frac{3}{4}=\frac{1\times4+3}{4}=\frac{7}{4}$

These two fractions are "unlike" fractions. (they have different denominators). Find equivalent fractions with the same denominator. Use this formula: $\frac{a}{b}-\frac{c}{d}=\frac{ad-bc}{bd}$

$\frac{19}{5}-\frac{7}{4}=\frac{(19)(4)-(5)(7)}{5\times4}=\frac{76-35}{20}=\frac{41}{20}$, the answer is an improper fraction, convert it into a mixed number. $\frac{41}{20}=2\frac{1}{20}$

2) Subtract. $4\frac{3}{8}-1\frac{1}{2}=$

Solution: Convert mixed numbers into fractions: $4\frac{3}{8}=\frac{4\times8+3}{8}=\frac{35}{8}$ and $1\frac{1}{2}=\frac{1\times2+1}{4}=\frac{3}{2}$

Find equivalent fractions: $\frac{3}{2}=\frac{12}{8}$. Then: $4\frac{3}{8}-1\frac{1}{2}=\frac{35}{8}-\frac{12}{8}=\frac{23}{8}$

The answer is an improper fraction, convert it into a mixed number.

$$\frac{23}{8}=2\frac{7}{8}$$

Multiplying Mixed Numbers

Use following steps for multiplying mixed numbers:

- ✓ Convert the mixed numbers into fractions. $a\frac{c}{b} = a + \frac{c}{b} = \frac{ab + c}{b}$
- ✓ Multiply fractions. $\frac{a}{b} \times \frac{c}{d} = \frac{a \times c}{b \times d}$
- ✓ Write your answer in lowest terms.
- ✓ If the answer is an improper fraction (numerator is bigger than denominator), convert it into a mixed number.

Examples:

1) Multiply. $3\frac{1}{3} \times 4\frac{1}{6} =$

Solution: Convert mixed numbers into fractions, $3\frac{1}{3} = \frac{3\times3+1}{3} = \frac{10}{3}$ and $4\frac{1}{6} = \frac{4\times6+1}{6} = \frac{25}{6}$

Apply the fractions rule for multiplication, $\frac{10}{3} \times \frac{25}{6} = \frac{10\times25}{3\times6} = \frac{250}{18}$

The answer is an improper fraction. Convert it into a mixed number. $\frac{250}{18} = 13\frac{8}{9}$

2) Multiply. $2\frac{1}{2} \times 3\frac{2}{3} =$

Solution: Converting mixed numbers into fractions, $2\frac{1}{2} \times 3\frac{2}{3} = \frac{5}{2} \times \frac{11}{3}$

Apply the fractions rule for multiplication, $\frac{5}{2} \times \frac{11}{3} = \frac{5\times11}{2\times3} = \frac{55}{6} = 9\frac{1}{6}$

3) Multiply mixed numbers. $2\frac{1}{3} \times 2\frac{1}{2} =$

Solution: Converting mixed numbers to fractions, $2\frac{1}{3} = \frac{7}{3}$ and $2\frac{1}{2} = \frac{5}{2}$. Multiply two fractions:

$$\frac{7}{3} \times \frac{5}{2} = \frac{7 \times 5}{3 \times 2} = \frac{35}{6} = 5\frac{5}{6}$$

Dividing Mixed Numbers

Use following steps for dividing mixed numbers:

- Convert the mixed numbers into fractions. $a\frac{c}{b} = a + \frac{c}{b} = \frac{ab + c}{b}$
- Divide fractions: Keep, Change, Flip: Keep first fraction, change division sign to multiplication, and flip the numerator and denominator of the second fraction. Then, solve! $\frac{a}{b} \div \frac{c}{d} = \frac{a}{b} \times \frac{d}{c} = \frac{a \times d}{b \times c}$
- Write your answer in lowest terms.
- If the answer is an improper fraction (numerator is bigger than denominator), convert it into a mixed number.

Examples:

1) Solve. $3\frac{2}{3} \div 2\frac{1}{2}$

Solution: Convert mixed numbers into fractions: $3\frac{2}{3} = \frac{3\times3+2}{3} = \frac{11}{3}$ and $2\frac{1}{2} = \frac{2\times2+1}{2} = \frac{5}{2}$

Keep, Change, Flip: $\frac{11}{3} \div \frac{5}{2} = \frac{11}{3} \times \frac{2}{5} = \frac{11\times2}{3\times5} = \frac{22}{15}$. The answer is an improper fraction. Convert it into a mixed number: $\frac{22}{15} = 1\frac{7}{15}$

2) Solve. $3\frac{4}{5} \div 1\frac{5}{6}$

Solution: Convert mixed numbers to fractions, then solve:

$$3\frac{4}{5} \div 1\frac{5}{6} = \frac{19}{5} \div \frac{11}{6} = \frac{19}{5} \times \frac{6}{11} = \frac{114}{55} = 2\frac{4}{55}$$

3) Solve. $2\frac{2}{7} \div 2\frac{3}{5}$

Solution: Converting mixed numbers to fractions: $3\frac{4}{5} \div 1\frac{5}{6} = \frac{16}{7} \div \frac{13}{5}$

Keep, Change, Flip: $\frac{16}{7} \div \frac{13}{5} = \frac{16}{7} \times \frac{5}{13} = \frac{16\times5}{7\times13} = \frac{80}{91}$

Chapter 1: Practices

Simplify each fraction.

1) $\frac{18}{30} =$

2) $\frac{21}{42} =$

3) $\frac{35}{55} =$

4) $\frac{48}{72} =$

5) $\frac{54}{81} =$

6) $\frac{80}{200} =$

Find the sum or difference.

7) $\frac{6}{15} + \frac{3}{15} =$

8) $\frac{2}{3} + \frac{1}{9} =$

9) $\frac{1}{4} + \frac{2}{5} =$

10) $\frac{7}{10} - \frac{3}{10} =$

11) $\frac{1}{2} - \frac{3}{8} =$

12) $\frac{5}{7} - \frac{3}{5} =$

Find the answers.

13) $\frac{1}{7} \div \frac{3}{8} =$

14) $\frac{2}{3} \times \frac{4}{7} =$

15) $\frac{5}{7} \times \frac{3}{4} =$

16) $\frac{2}{5} \div \frac{3}{7} =$

17) $\frac{3}{7} \div \frac{5}{8} =$

18) $\frac{3}{8} \times \frac{4}{7} =$

Calculate.

19) $3\frac{1}{5} + 2\frac{2}{9} =$

20) $1\frac{1}{7} + 5\frac{2}{5} =$

21) $4\frac{4}{5} + 1\frac{2}{7} =$

22) $2\frac{4}{7} + 2\frac{3}{5} =$

23) $1\frac{5}{6} + 1\frac{2}{5} =$

24) $3\frac{5}{7} + 1\frac{2}{9} =$

Calculate.

25) $3\frac{2}{5} - 1\frac{2}{9} =$

26) $5\frac{3}{5} - 1\frac{1}{7} =$

27) $4\frac{2}{5} - 2\frac{2}{7} =$

28) $8\frac{3}{4} - 2\frac{1}{8} =$

29) $9\frac{5}{7} - 7\frac{4}{21} =$

30) $11\frac{7}{12} - 9\frac{5}{6} =$

Find the answers.

31) $1\frac{1}{8} \times 1\frac{3}{4} =$

32) $3\frac{1}{5} \times 2\frac{2}{7} =$

33) $2\frac{1}{8} \times 1\frac{2}{9} =$

34) $2\frac{3}{8} \times 2\frac{2}{5} =$

35) $1\frac{1}{2} \times 5\frac{2}{3} =$

36) $3\frac{1}{2} \times 6\frac{2}{3} =$

Solve.

37) $9\frac{1}{2} \div 2\frac{3}{5} =$

38) $2\frac{3}{8} \div 1\frac{2}{5} =$

39) $5\frac{3}{4} \div 2\frac{2}{7} =$

40) $8\frac{1}{3} \div 4\frac{1}{4} =$

41) $7\frac{2}{5} \div 3\frac{3}{4} =$

42) $2\frac{4}{5} \div 3\frac{2}{3} =$

Answers – Chapter 1

1) $\frac{3}{5}$
2) $\frac{1}{2}$
3) $\frac{7}{11}$
4) $\frac{2}{3}$
5) $\frac{2}{3}$
6) $\frac{2}{5}$
7) $\frac{3}{5}$
8) $\frac{7}{9}$
9) $\frac{13}{20}$
10) $\frac{2}{5}$
11) $\frac{1}{8}$
12) $\frac{4}{35}$
13) $\frac{8}{21}$
14) $\frac{8}{21}$
15) $\frac{15}{28}$
16) $\frac{14}{15}$
17) $\frac{24}{35}$
18) $\frac{3}{14}$
19) $5\frac{19}{45}$
20) $6\frac{19}{35}$
21) $6\frac{3}{35}$
22) $5\frac{6}{35}$
23) $3\frac{7}{30}$
24) $4\frac{59}{63}$
25) $2\frac{8}{45}$
26) $6\frac{16}{35}$
27) $2\frac{4}{35}$
28) $6\frac{5}{8}$
29) $2\frac{11}{21}$
30) $1\frac{3}{4}$
31) $1\frac{31}{32}$
32) $7\frac{11}{35}$
33) $2\frac{43}{72}$
34) $5\frac{7}{10}$
35) $8\frac{1}{2}$
36) $23\frac{1}{3}$
37) $3\frac{17}{26}$
38) $1\frac{39}{56}$
39) $2\frac{33}{64}$
40) $1\frac{49}{51}$
41) $1\frac{73}{75}$
42) $\frac{42}{55}$

Chapter 2:

Decimals

Math Topics that you'll learn in this Chapter:

- ✓ Comparing Decimals
- ✓ Rounding Decimals
- ✓ Adding and Subtracting Decimals
- ✓ Multiplying and Dividing Decimals

Comparing Decimals

- Decimal is a fraction written in a special form. For example, instead of writing $\frac{1}{2}$ you can write $\mathbf{0.5}$
- A Decimal Number contains a Decimal Point. It separates the whole number part from the fractional part of a decimal number.
- Let's review decimal place values: Example: $\mathbf{53.9861}$

5: tens	3: ones	9: tenths
8: hundredths	6: thousandths	1: tens thousandths

✓ To compare decimals, compare each digit of two decimals in the same place value. Start from left. Compare hundreds, tens, ones, tenth, hundredth, etc.

✓ To compare numbers, use these symbols:

Equal to $=$, Less than $<$, Greater than $>$
Greater than or equal $\geq$, Less than or equal $\leq$

Examples:

1) Compare 0.60 and 0.06.

Solution: 0.60 *is greater than* 0.06, because the tenth place of 0.60 is 6, but the tenth place of 0.06 is zero. Then: $0.60 > 0.06$

2) Compare 0.0815 and 0.815.

Solution: 0.815 *is greater than* 0.0815, because the tenth place of 0.815 is 8, but the tenth place of 0.0815 is zero. Then: $0.0815 < 0.815$

Rounding Decimals

- We can round decimals to a certain accuracy or number of decimal places. This is used to make calculation easier to do and results easier to understand, when exact values are not too important.
- First, you'll need to remember your place values: For example:

$$12.4869$$

1: tens	2: ones	4: tenths
8: hundredths	6: thousandths	9: tens thousandths

- To round a decimal, first find the place value you'll round to.
- Find the digit to the right of the place value you're rounding to. If it is 5 or bigger, add 1 to the place value you're rounding to and remove all digits on its right side. If the digit to the right of the place value is less than 5, keep the place value and remove all digits on the right.

Examples:

1) Round 1.9278 to the thousandth place value.

Solution: First look at the next place value to the right, (tens thousandths). It's 8 and it is greater than 5. Thus add 1 to the digit in the thousandth place. Thousandth place is 7. $\rightarrow$ $7 + 1 = 8$, then, the answer is 1.928

2) Round 9.4126 to the nearest hundredth.

Solution: First look at the digit to the right of hundredth (thousandths place value). It's 2 and it is less than 5, thus remove all the digits to the right of hundredth place. Then, the answer is 9.41

Adding and Subtracting Decimals

- ✓ Line up the decimal numbers.
- ✓ Add zeros to have same number of digits for both numbers if necessary.
- ✓ Remember your place values: For example:

73.5196

7: tens	3: ones	5: tenths
1: hundredths	9: thousandths	6: tens thousandths

- ✓ Add or subtract using column addition or subtraction.

Examples:

1) Add. $1.8 + 3.12$

Solution: First line up the numbers: $\begin{array}{r} 1.8 \\ +3.12 \\ \hline \end{array}$ → Add a zero to have same number of digits for both numbers. $\begin{array}{r} 1.80 \\ +3.12 \\ \hline \end{array}$ → Start with the hundredths place: $0 + 2 = 2$, $\begin{array}{r} 1.80 \\ +3.12 \\ \hline 2 \end{array}$ → Continue with tenths place: $8 + 1 = 9$, $\begin{array}{r} 1.80 \\ +3.12 \\ \hline .92 \end{array}$ → Add the ones place: $3 + 1 = 4$, $\begin{array}{r} 1.80 \\ +3.12 \\ \hline 4.92 \end{array}$

2) Find the difference. $3.67 - 2.23$

Solution: First line up the numbers: $\begin{array}{r} 3.67 \\ -2.23 \\ \hline \end{array}$ → Start with the hundredths place: $7 - 3 = 4$, $\begin{array}{r} 3.67 \\ -2.23 \\ \hline 4 \end{array}$ → Continue with tenths place. $6 - 2 = 4$, $\begin{array}{r} 3.67 \\ -2.23 \\ \hline .44 \end{array}$ → Subtract the ones place. $3 - 2 = 1$, $\begin{array}{r} 3.67 \\ -2.23 \\ \hline 1.44 \end{array}$

Multiplying and Dividing Decimals

For multiplying decimals:

✓ Ignore the decimal point and set up and multiply the numbers as you do with whole numbers.

✓ Count the total number of decimal places in both of the factors.

✓ Place the decimal point in the product.

For dividing decimals:

✓ If the divisor is not a whole number, move decimal point to right to make it a whole number. Do the same for dividend.

✓ Divide similar to whole numbers.

Examples:

1) Find the product. $0.81 \times 0.32 =$

Solution: Set up and multiply the numbers as you do with whole numbers. Line up the numbers: $\frac{81}{\times 32}$ → Start with the ones place then continue with other digits → $\frac{\begin{array}{r}81\\ \times 32\end{array}}{2{,}592}$. Count the total number of decimal places in both of the factors. There are four decimals digits. (two for each factor 0.81 and 0.32) Then: $0.81 \times 0.32 = 0.2592$

2) Find the quotient. $1.60 \div 0.4 =$

Solution: The divisor is not a whole number. Multiply it by 10 to get 4: → $0.4 \times 10 = 4$

Do the same for the dividend to get 16. → $1.60 \times 10 = 1.6$

Now, divide: $16 \div 4 = 4$. The answer is 4.

Chapter 2: Practices

Compare. Use $>$, $=$, and $<$

1) $0.88 \square 0.088$
2) $0.56 \square 0.57$
3) $0.99 \square 0.89$
4) $1.55 \square 1.65$
5) $1.58 \square 1.75$
6) $2.91 \square 2.85$

Round each decimal to the nearest whole number.

7) 5.94
8) 16.47
9) 9.7
10) 35.8
11) 24.46
12) 12.5

Find the sum or difference.

13) $43.15 + 23.65 =$
14) $56.74 - 22.43 =$
15) $25.47 + 31.76 =$
16) $69.87 - 35.98 =$
17) $45.53 + 18.95 =$
18) $25.13 - 18.72 =$

Find the product and quotient.

19) $0.5 \times 0.8 =$
20) $6.4 \div 0.4 =$
21) $3.25 \times 2.2 =$
22) $8.4 \div 2.5 =$
23) $5.4 \times 0.6 =$
24) $1.42 \div 0.5 =$

Answers – Chapter 2

1) $0.88 > 0.088$
2) $0.56 < 0.57$
3) $0.99 > 0.89$
4) $1.55 < 1.65$
5) $1.58 < 1.75$
6) $2.91 > 2.85$
7) 6
8) 16
9) 10
10) 36
11) 24
12) 13
13) 66.8
14) 34.31
15) 57.23
16) 33.89
17) 64.48
18) 6.41
19) 0.4
20) 16
21) 7.15
22) 3.36
23) 3.24
24) 2.84

Chapter 3:

Integers and Order of Operations

Math Topics that you'll learn in this Chapter:

- ✓ Adding and Subtracting Integers
- ✓ Multiplying and Dividing Integers
- ✓ Order of Operations
- ✓ Integers and Absolute Value

Adding and Subtracting Integers

- Integers include: zero, counting numbers, and the negative of the counting numbers. $\{\ldots, -3, -2, -1, 0, 1, 2, 3, \ldots\}$
- Add a positive integer by moving to the right on the number line. (you will get a bigger number)
- Add a negative integer by moving to the left on the number line. (you will get a smaller number)
- Subtract an integer by adding its opposite.

Examples:

1) Solve. $(-4) - (-5) =$

Solution: Keep the first number and convert the sign of the second number to its opposite. (change subtraction into addition. Then: $(-4) + 5 = 1$

2) Solve. $11 + (8 - 19) =$

Solution: First subtract the numbers in brackets, $8 - 19 = -11$.

Then: $11 + (-11) =$ → change addition into subtraction: $11 - 11 = 0$

3) Solve. $5 - (-14 - 3) =$

Solution: First subtract the numbers in brackets, $-14 - 3 = -17$

Then: $5 - (-17) =$ → change subtraction into addition: $5 + 17 = 22$

4) Solve. $10 + (-6 - 15) =$

Solution: First subtract the numbers in brackets, $-6 - 15 = -21$

Then: $10 + (-21) =$ → change addition into subtraction: $10 - 21 = -11$

Multiplying and Dividing Integers

Use following rules for multiplying and dividing integers:

- ✓ (negative) × (negative) = positive
- ✓ (negative) ÷ (negative) = positive
- ✓ (negative) × (positive) = negative
- ✓ (negative) ÷ (positive) = negative
- ✓ (positive) × (positive) = positive
- ✓ (positive) ÷ (negative) = negative

Examples:

1) Solve. $2 \times (-3) =$

Solution: Use this rule: (positive) × (negative) = negative.
Then: $(2) \times (-3) = -6$

2) Solve. $(-5) + (-27 \div 9) =$

Solution: First divided -27 by 9 , the numbers in brackets, use this rule: (negative) ÷ (positive) = negative. Then: $-27 \div 9 = -3$
$(-5) + (-27 \div 9) = (-5) + (-3) = -5 - 3 = -8$

3) Solve. $(15 - 17) \times (-8) =$

Solution: First subtract the numbers in brackets, $15 - 17 = -2 \rightarrow (-2) \times (-8) =$

Now use this rule: (negative) × (negative) = positive
$(-2) \times (-8) = 16$

4) Solve. $(16 - 10) \div (-2) =$

Solution: First subtract the numbers in brackets, $16 - 10 = 6 \rightarrow (6) \div (-2) =$

Now use this rule: (positive) ÷ (negative) = negative
$(6) \div (-2) = -3$

Order of Operations

- In Mathematics, "operations" are addition, subtraction, multiplication, division, exponentiation (written as b^n), and grouping;
- When there is more than one math operation in an expression, use PEMDAS: (to memorize this rule, remember the phrase "Please Excuse My Dear Aunt Sally".)
  - Parentheses
  - Exponents
  - Multiplication and Division (from left to right)
  - Addition and Subtraction (from left to right)

Examples:

1) Calculate. $(3+5) \div (3^2 \div 9) =$

   Solution: First simplify inside parentheses: $(8) \div (9 \div 9) = (8) \div (1)$, Then: $(8) \div (1) = 8$

2) Solve. $(7 \times 8) - (12 - 4) =$

   Solution: First calculate within parentheses: $(7 \times 8) - (12 - 4) = (56) - (8)$, Then: $(56) - (8) = 48$

3) Calculate. $-2[(8 \times 9) \div (2^2 \times 2)] =$

   Solution: First calculate within parentheses: $-2[(72) \div (4 \times 2)] = -2[(72) \div (8)] = -2[9]$ multiply -2 and 9. Then: $-2[9] = -18$

4) Solve. $(14 \div 7) + (-13 + 8) =$

   Solution: First calculate within parentheses: $(14 \div 7) + (-13 + 8) = (2) + (-5)$

   Then: $(2) - (5) = -3$

Integers and Absolute Value

- The absolute value of a number is its distance from zero, in either direction, on the number line. For example, the distance of 9 and -9 from zero on number line is 9.
- The absolute value of an integer is the numerical value without its sign. (negative or positive)
- The vertical bar is used for absolute value as in $|x|$.
- The absolute value of a number is never negative; because it only shows, "how far the number is from zero".

Examples:

1) Calculate. $|12-4| \times 4 =$

Solution: First solve $|12-4|$, →$|12-4| = |8|$, the absolute value of 8 is 8, $|8| = 8$
Then: $8 \times 4 = 32$

2) Solve. $\frac{|-16|}{4} \times |3-8| =$

Solution: First find $|-16|$, → the absolute value of -16 is 16, then: $|-16| = 16$,
$\frac{16}{4} \times |3-8| =$
Now, calculate $|3-8|$, → $|3-8| = |-5|$, the absolute value of -5 is 5. $|-5| = 5$
Then: $\frac{16}{4} \times 5 = 4 \times 5 = 20$

3) Solve. $|9-3| \times \frac{|-3\times 8|}{6} =$

Solution: First calculate $|9-3|$, →$|9-3| = |6|$, the absolute value of 6 is 6, $|6| = 6$. Then:
$6 \times \frac{|-3\times 8|}{6}$
Now calculate $|-3 \times 8|$, → $|-3 \times 8| = |-24|$, the absolute value of -24 is 24, $|-24| = 24$
Then: $6 \times \frac{24}{6} = 6 \times 4 = 24$

Chapter 3: Practices

Find each sum or difference.

1) $18 + (-5) =$
2) $(-16) + 24 =$
3) $(-12) + (-9) =$
4) $14 + (-8) + 6 =$
5) $24 + (-10 - 7) =$
6) $(-15) + (-6 + 12) =$

Find each product or quotient.

7) $8 \times (-6) =$
8) $(-12) \div (-3) =$
9) $(-4) \times (-7) \times 2 =$
10) $3 \times (-5) \times (-6) =$
11) $(-7 - 37) \div (-11) =$
12) $(8 - 6) \times (-24) =$

Evaluate each expression.

13) $8 + (3 \times 7) =$
14) $(18 \times 2) - 14 =$
15) $(15 - 7) + (2 \times 6) =$
16) $(8 + 4) \div (2^3 \div 2) =$
17) $2[(6 \times 3) \div (3^2 \times 2)] =$
18) $-3[(8 \times 2^2) \div (8 \times 2)] =$

Find the answers.

19) $|-6| + |9 - 12| =$
20) $|8| - |7 - 19| + 1 =$
21) $\frac{|-40|}{8} \times \frac{|-15|}{5} =$
22) $|7 \times -5| \times \frac{|-32|}{8} =$
23) $\frac{|-121|}{11} - |-8 \times 2| =$
24) $\frac{|-3 \times -6|}{9} \times \frac{|4 \times -6|}{8} =$

Answers – Chapter 3

1) 13
2) 8
3) -21
4) 12
5) 7
6) -9
7) -48
8) 4
9) 56
10) 90
11) 4
12) -48
13) 29
14) 22
15) 20
16) 3
17) 2
18) -6
19) 9
20) -3
21) 15
22) 140
23) -5
24) 6

Chapter 4:

Ratios and Proportions

Math Topics that you'll learn in this Chapter:

- ✓ Simplifying Ratios
- ✓ Proportional Ratios
- ✓ Similarity and Ratios

Simplifying Ratios

- Ratios are used to make comparisons between two numbers.
- Ratios can be written as a fraction, using the word "to", or with a colon. Example: $\frac{3}{4}$ or "3 to 4" or 3:4
- You can calculate equivalent ratios by multiplying or dividing both sides of the ratio by the same number.

1) Simplify. $9:3 =$

Solution: Both numbers 9 and 3 are divisible by 3 , $\Rightarrow 9 \div 3 = 3$, $3 \div 3 = 1$, Then: $9:3 = 3:1$

2) Simplify. $\frac{24}{44} =$

Solution: Both numbers 24 and 44 are divisible by 4, $\Rightarrow$ $24 \div 4 = 6$, $44 \div 4 = 11$, Then: $\frac{24}{44} = \frac{6}{11}$

3) There are 36 students in a class and 16 of them are girls. Write the ratio of girls to boys.

Solution: Subtract 16 from 36 to find the number of boys in the class. $36 - 16 = 20$. There are 20 boys in the class. So, ratio of girls to boys is $16:20$. Now, simplify this ratio. Both 20 and 16 are divisible by 4. Then: $20 \div 4 = 5$, and $16 \div 4 = 4$. In simplest form, this ratio is $4:5$

4) A recipe calls for butter and sugar in the ratio $3:4$. If you're using 9 cups of butter, how many cups of sugar should you use?

Solution: Since, you use 9 cups of butter, or 3 times as much, you need to multiply the amount of sugar by 3. Then: $4 \times 3 = 12$. So, you need to use 12 cups of sugar. You can solve this using equivalent fractions: $\frac{3}{4} = \frac{9}{12}$

Proportional Ratios

- Two ratios are proportional if they represent the same relationship.
- A proportion means that two ratios are equal. It can be written in two ways: $\frac{a}{b} = \frac{c}{d}$ $\quad a : b = c : d$
- The proportion $\frac{a}{b} = \frac{c}{d}$ can be written as: $a \times d = c \times b$

Examples:

1) Solve this proportion for x. $\frac{3}{7} = \frac{12}{x}$

Solution: Use cross multiplication: $\frac{3}{7} = \frac{12}{x} \Rightarrow 3 \times x = 7 \times 12 \Rightarrow 3x = 84$

Divide both sides by 3 to find x: $x = \frac{84}{3} \Rightarrow x = 28$

2) If a box contains red and blue balls in ratio of $3: 7$ red to blue, how many red balls are there if 49 blue balls are in the box?

Solution: Write a proportion and solve. $\frac{3}{7} = \frac{x}{49}$

Use cross multiplication: $3 \times 49 = 7 \times x \Rightarrow 147 = 7x$

Divide to find x: $x = \frac{147}{7} \Rightarrow x = 21$. There are 21 red balls in the box.

3) Solve this proportion for x. $\frac{2}{9} = \frac{12}{x}$

Solution: Use cross multiplication: $\frac{2}{9} = \frac{12}{x} \Rightarrow 2 \times x = 9 \times 12 \Rightarrow 2x = 108$

Divide to find x: $x = \frac{108}{2} \Rightarrow x = 54$

4) Solve this proportion for x. $\frac{6}{7} = \frac{18}{x}$

Solution: Use cross multiplication: $\frac{6}{7} = \frac{18}{x} \Rightarrow 6 \times x = 7 \times 18 \Rightarrow 6x = 126$

Divide to find x: $x = \frac{126}{6} \Rightarrow x = 21$

Similarity and Ratios

- Two figures are similar if they have the same shape.
- Two or more figures are similar if the corresponding angles are equal, and the corresponding sides are in proportion.

Examples:

1) Following triangles are similar. What is the value of unknown side?

Solution: Find the corresponding sides and write a proportion.

$\frac{5}{10} = \frac{4}{x}$. Now, use cross product to solve for x:

$\frac{5}{10} = \frac{4}{x} \rightarrow 5 \times x = 10 \times 4 \rightarrow 5x = 40$. Divide both sides by 5. Then: $5x = 40 \rightarrow \frac{5x}{5} = \frac{40}{5} \rightarrow x = 8$

The missing side is 8.

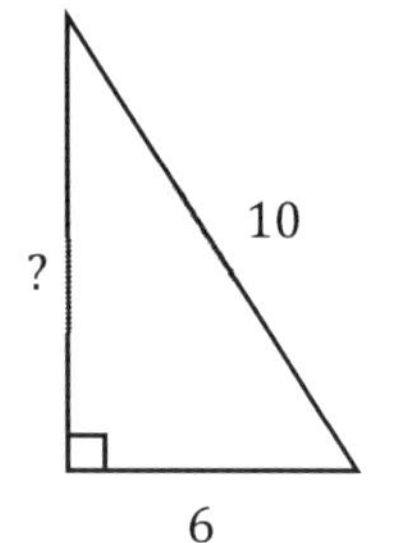

2) Two rectangles are similar. The first is 6 feet wide and 20 feet long. The second is 15 feet wide. What is the length of the second rectangle?

Solution: Let's put x for the length of the second rectangle. Since two rectangles are similar, their corresponding sides are in proportion. Write a proportion and solve for the missing number. $\frac{6}{15} = \frac{20}{x} \rightarrow 6x = 15 \times 20 \rightarrow 6x = 300 \rightarrow x = \frac{300}{6} = 50$

The length of the second rectangle is 50 feet.

Chapter 4: Practices

Reduce each ratio.

1) $9:18 =$ ___:___
2) $6:54 =$ ___:___
3) $28:49 =$ ___:___
4) Bob has 12 red cards and 20 green cards. What is the ratio of Bob's red cards to his green cards? ________________

5) In a party, 10 soft drinks are required for every 12 guests. If there are 252 guests, how many soft drinks is required? ________________

6) In Jack's class, 18 of the students are tall and 10 are short. In Michael's class 54 students are tall and 30 students are short. Which class has a higher ratio of tall to short students? ________________

Solve each proposition.

7) $\frac{3}{7} = \frac{18}{x}, x =$ ____
8) $\frac{5}{9} = \frac{x}{108}, x =$ ____
9) $\frac{2}{13} = \frac{8}{x}, x =$ ____
10) $\frac{4}{10} = \frac{6}{x}, x =$ ____
11) $\frac{8}{20} = \frac{x}{65}, x =$ ____
12) $\frac{6}{15} = \frac{14}{x}, x =$ ____

Solve each problem.

13) Two rectangles are similar. The first is $8\,feet$ wide and $22\,feet$ long. The second is $12\,feet$ wide. What is the length of the second rectangle? ________________

14) Two rectangles are similar. One is $3.2\,meters$ by $8\,meters$. The longer side of the second rectangle is $34.5\,meters$. What is the other side of the second rectangle? ________________

Answers – Chapter 4

1) 1: 2
2) 1: 9
3) 4: 7
4) 3: 5
5) 210
6) The ratio for both classes is 9 to 5.
7) 42
8) 60
9) 52
10) 15
11) 26
12) 35
13) 33 feet
14) 13.8 meters

Chapter 5:

Percentage

Math Topics that you'll learn in this Chapter:

- ✓ Percentage Calculations
- ✓ Percent Problems
- ✓ Percent of Increase and Decrease
- ✓ Discount, Tax and Tip
- ✓ Simple Interest

Percent Problems

- Percent is a ratio of a number and 100. It always has the same denominator, 100. Percent symbol is "%".
- Percent means "per 100". So, 20% is 20/100.
- In each percent problem, we are looking for the base, or part or the percent.
- Use the following equations to find each missing section in a percent problem:
  - Base = Part ÷ Percent
  - Part = Percent × Base
  - Percent = Part ÷ Base

Examples:

1) What is 25% of 60?

Solution: In this problem, we have percent (25%) and base (60) and we are looking for the "part".

Use this formula: $part = percent \times base$. Then: $part = 25\% \times 60 = \frac{25}{100} \times 60 = 0.25 \times 60 = 15$.

The answer:

25% of 60 is 15.

2) 20 is what percent of 400?

Solution: In this problem, we are looking for the percent. Use this equation:

$Percent = Part \div Base \rightarrow Percent = 20 \div 400 = 0.05 = 5\%$.

Then: 20 is 5 percent of 400.

Percent of Increase and Decrease

- Percent of change (increase or decrease) is a mathematical concept that represents the degree of change over time.
- To find the percentage of increase or decrease:

1. New Number - Original Number
2. The result ÷ Original Number × 100

- Or use this formula: Percent of change = $\frac{new\ number - original\ number}{original\ number} \times 100$
- Note: If your answer is a negative number, then this is a percentage decrease. If it is positive, then this is a percentage increase.

Examples:

1) The price of a shirt increases from $20 to $30. What is the percentage increase?
Solution: First find the difference: $30 - 20 = 10$
Then: $10 \div 20 \times 100 = \frac{10}{20} \times 100 = 50$. The percentage increase is 50. It means that the price of the shirt increased 50%.

2) The price of a table increased from $25 to $40. What is the percent of increase?
Solution: Use percentage formula: $Percent\ of\ change = \frac{new\ number - original\ number}{original\ number} \times 100 = \frac{40-25}{25} \times 100 = \frac{15}{25} \times 100 = 0.6 \times 100 = 60$. The percentage increase is 60. It means that the price of the table increased 60%.

3) The population of a town was 50,000 in the 2000 census and 40,000 in the 2010 census. By what percent did the population decrease?
Solution: Use percentage formula:
$Percent\ of\ change = \frac{new\ number - original\ number}{original\ number} \times 100 = \frac{40{,}000 - 50{,}000}{50{,}000} \times 100 = \frac{-10{,}000}{50{,}000} \times 100 = -0.2 \times 100 = -20$. The population of the town decreased by 20%.

Discount, Tax and Tip

- To find discount: Multiply the regular price by the rate of discount
- To find selling price: Original price - discount
- To find tax: Multiply the tax rate to the taxable amount (income, property value, etc.)
- To find tip, multiply the rate to the selling price.

Examples:

1) With an 10% discount, Ella was able to save $\$45$ on a dress. What was the original price of the dress?

   Solution: let x be the original price of the dress. Then: $10\ \%\ of\ x = 45$. Write an equation and solve for x: $0.10 \times x = 45 \rightarrow x = \frac{45}{0.10} = 450$. The original price of the dress was $450.

2) Sophia purchased a new computer for a price of $950 at the Apple Store. What is the total amount her credit card is charged if the sales tax is 7%?

   Solution: The taxable amount is $950, and the tax rate is 7%. Then: $Tax = 0.07 \times 950 = 66.50$

   $Final\ price = Selling\ price + Tax \rightarrow final\ price = \$950 + \$66.50 = \$1{,}016.50$

3) Nicole and her friends went out to eat at a restaurant. If their bill was $80.00 and they gave their server a 15% tip, how much did they pay altogether?

   Solution: First find the tip. To find tip, multiply the rate to the bill amount.

   $Tip = 80 \times 0.15 = 12$. The final price is: $\$80 + \$12 = \$92$

Simple Interest

- Simple Interest: The charge for borrowing money or the return for lending it.
- Simple interest is calculated on the initial amount (principal).
- To solve a simple interest problem, use this formula:

  Interest = principal x rate x time $(\boldsymbol{I = p \times r \times t = prt})$

Examples:

1) Find simple interest for $\$300$ investment at 6% for 5 years.

   Solution: Use Interest formula: $I = prt$ ($P = \$300$, $\text{r} = 6\% = \frac{6}{100} = 0.06$ and $t = 5$)

   Then: $I = 300 \times 0.06 \times 5 = \$90$

2) Find simple interest for $\$1{,}600$ at 5% for 2 years.

   Solution: Use Interest formula: $I = prt$ ($P = \$1{,}600$, $\text{r} = 5\% = \frac{5}{100} = 0.05$ and $t = 2$)

   Then: $I = 1{,}600 \times 0.05 \times 2 = \$160$

3) Andy received a student loan to pay for his educational expenses this year. What is the interest on the loan if he borrowed $6,500 at 8% for 6 years?

   Solution: Use Interest formula: $I = prt$. $P = \$6{,}500$, $\text{r} = 8\% = 0.08$ and $t = 6$

   Then: $I = 6{,}500 \times 0.08 \times 8 = \$3{,}120$

4) Bob is starting his own small business. He borrowed $10,000 from the bank at a 6% rate for 6 months. Find the interest Bob will pay on this loan.

   Solution: Use Interest formula: $I = prt$. $P = \$10{,}000$, $\text{r} = 6\% = 0.06$ and $t = 0.5$ (6 months is half year). Then: $I = 10{,}000 \times 0.06 \times 0.5 = \$300$

Chapter 5: Practices

Solve each problem.

1) 15 is what percent of 60? ____%
2) 18 is what percent of 24? ____%
3) 25 is what percent of 500? ____%
4) 14 is what percent of 280? ____%
5) 45 is what percent of 180? ____%
6) 70 is what percent of 350? ____%

Solve each percent of change word problem.

7) Bob got a raise, and his hourly wage increased from $15 to $18. What is the percent increase? ______ %
8) The price of a pair of shoes increases from $40 to $66. What is the percent increase? ____ %
9) At a coffeeshop, the price of a cup of coffee increased from $1.20 to $1.44. What is the percent increase in the cost of the coffee? ______ %

Find the selling price of each item.

10)Original price of a computer: $650

Tax: 8%, Selling price: $________

11) Original price of a laptop: $480

Tax: 15%, Selling price: $________

12) Nicolas hired a moving company. The company charged $400 for its services, and Nicolas gives the movers a 15% tip. How much does Nicolas tip the movers? $__________

13) Mason has lunch at a restaurant and the cost of his meal is $30. Mason wants to leave a 20% tip. What is Mason's total bill including tip? $__________

Determine the simple interest for these loans.

14) $\$440\ at\ 5\%\ for\ 6\ years.$ $___

15) $\$460\ at\ 2.5\%\ for\ 4\ years.$ $_

16) $\$500\ at\ 3\%\ for\ 5\ years.$ $___

17) $550 at 9% for 2 years. $___

18) A new car, valued at $28,000, depreciates at 9% per year. What is the value of the car one year after purchase? $______________

19) Sara puts $4,000 into an investment yielding 5% annual simple interest; she left the money in for five years. How much interest does Sara get at the end of those five years? $______________

Answers – Chapter 5

1) 25%
2) 75%
3) 5%
4) 5%
5) 25%
6) 20%
7) 20%
8) 65%
9) 20%
10) $702.00
11) $552.00
12) $60.00
13) $36.00
14) $132
15) $46
16) $75
17) $99
18) $25,480.00
19) $1,000.00

Chapter 6:

Expressions and Variables

Math Topics that you'll learn in this Chapter:

- ✓ Simplifying Variable Expressions
- ✓ Simplifying Polynomial Expressions
- ✓ The Distributive Property
- ✓ Evaluating One Variable
- ✓ Evaluating Two Variables

Simplifying Variable Expressions

- In algebra, a variable is a letter used to stand for a number. The most common letters are: $x, y, z, a, b, c, m, and\ n$.
- Algebraic expression is an expression contains integers, variables, and the math operations such as addition, subtraction, multiplication, division, etc.
- In an expression, we can combine "like" terms. (values with same variable and same power)

Examples:

1) Simplify. $(2x + 3x + 4) =$

Solution: In this expression, there are three terms: $2x, 3x$, and 4. Two terms are "like terms": $2x$ and $3x$. Combine like terms. $2x + 3x = 5x$. Then: $(2x + 3x + 4) = 5x + 4$ (remember you cannot combine variables and numbers.)

2) Simplify. $12 - 3x^2 + 5x + 4x^2 =$

Solution: Combine "like" terms: $-3x^2 + 4x^2 = x^2$. Then:
$12 - 3x^2 + 5x + 4x^2 = 12 + x^2 + 5x$. Write in standard form (biggest powers first):
$12 + x^2 + 5x = x^2 + 5x + 12$

3) Simplify. $(10x^2 + 2x^2 + 3x) =$

Solution: Combine like terms. Then: $(10x^2 + 2x^2 + 3x) = 12x^2 + 3x$

4) Simplify. $15x - 3x^2 + 9x + 5x^2 =$

Solution: Combine "like" terms: $15x + 9x = 24x$, and $-3x^2 + 5x^2 = 2x^2$

Then: $15x - 3x^2 + 9x + 5x^2 = 24x + 2x^2$. Write in standard form (biggest powers first): $24x + 2x^2 = 2x^2 + 24x$

Simplifying Polynomial Expressions

- In mathematics, a polynomial is an expression consisting of variables and coefficients that involves only the operations of addition, subtraction, multiplication, and non-negative integer exponents of variables. $P(x) = a_n x^n + a_{n-1} x^{n-1} + \dots + a_2 x^2 + a_1 x + a_0$
- Polynomials must always be simplified as much as possible. It means you must add together any like terms. (values with same variable and same power)

Examples:

1) Simplify this Polynomial Expressions. $x^2 - 5x^3 + 2x^4 - 4x^3$

   Solution: Combine "like" terms: $-5x^3 - 4x^3 = -9x^3$

   Then: $x^2 - 5x^3 + 2x^4 - 4x^3 = x^2 - 9x^3 + 2x^4$

   Now, write the expression in standard form: $2x^4 - 9x^3 + x^2$

2) Simplify this expression. $(2x^2 - x^3) - (x^3 - 4x^2) =$

   Solution: First use distributive property: → multiply $(-)$ into $(x^3 - 4x^2)$

   $(2x^2 - x^3) - (x^3 - 4x^2) = 2x^2 - x^3 - x^3 + 4x^2$

   Then combine "like" terms: $2x^2 - x^3 - x^3 + 4x^2 = 6x^2 - 2x^3$

   And write in standard form: $6x^2 - 2x^3 = -2x^3 + 6x^2$

3) Simplify. $4x^4 - 5x^3 + 15x^4 - 12x^3 =$

   Solution: Combine "like" terms: $-5x^3 - 12x^3 = -17x^3$ and $4x^4 + 15x^4 = 19x^4$

   Then: $4x^4 - 5x^3 + 15x^4 - 12x^3 = 19x^4 - 17x^3$

The Distributive Property

- The distributive property (or the distributive property of multiplication over addition and subtraction) simplifies and solves expressions in the form of: $a(b+c)$ or $a(b-c)$
- The distributive property is multiplying a term outside the parentheses by the terms inside.
- Distributive Property rule: $a(b+c) = ab + ac$

Examples:

1) *Simply using distributive property.* $(-4)(x-5)$

Solution: Use Distributive Property rule: $a(b+c) = ab + ac$

$(-4)(x-5) = (-4 \times x) + (-4) \times (-5) = -4x + 20$

2) *Simply.* $(3)(2x-4)$

Solution: Use Distributive Property rule: $a(b+c) = ab + ac$

$(3)(2x-4) = (3 \times 2x) + (3) \times (-4) = 6x - 12$

3) *Simply.* $(-3)(3x-5) + 4x$

Solution: First, simplify $(-3)(3x-5)$ using distributive property.

Then: $(-3)(3x-5) = -9x + 15$

Now combine like terms: $(-3)(3x-5) + 4x = -9x + 15 + 4x$

In this expression, $-9x$ and $4x$ are "like terms" and we can combine them.

$-9x + 4x = -5x$. Then: $-9x + 15 + 4x = -5x + 15$

Evaluating One Variable

- To evaluate one variable expressions, find the variable and substitute a number for that variable.
- Perform the arithmetic operations.

Examples:

1) *Calculate this expression for* $\mathrm{x} = 3$. $15 - 3x$

Solution: First substitute 3 for x

Then: $15 - 3x = 15 - 3(3)$

Now, use order of operation to find the answer: $15 - 3(3) = 15 - 9 = 6$

2) *Evaluate this expression for* $\mathrm{x} = 1$. $5x - 12$

Solution: First substitute 1 for x, then:

$5x - 12 = 5(1) - 12$

Now, use order of operation to find the answer: $5(1) - 12 = 5 - 12 = -7$

3) *Find the value of this expression when* $\mathrm{x} = 5$. $25 - 4x$

Solution: First substitute 5 for x, then:

$25 - 4x = 25 - 4(5) = 25 - 20 = 5$

4) *Solve this expression for* $x = -2$. $12 + 3x$

Solution: Substitute -2 for x, then: $12 + 3x = 12 + 3(-2) = 12 - 6 = 6$

Evaluating Two Variables

- To evaluate an algebraic expression, substitute a number for each variable.
- Perform the arithmetic operations to find the value of the expression.

Examples:

1) *Calculate this expression for* $a = 3$ *and* $b = -2$. $3a - 6b$

Solution: First substitute 3 for a, and -2 for b, then:

$$3a - 6b = 3(3) - 6(-2)$$

Now, use order of operation to find the answer: $3(3) - 6(-2) = 9 + 12 = 21$

2) *Evaluate this expression for* $x = 3$ *and* $y = 1$. $3x + 5y$

Solution: Substitute 3 for x, and 1 for y, then:

$$3x + 5y = 3(3) + 5(1) = 9 + 5 = 14$$

3) *Find the value of this expression when* $a = 1$ *and* $b = 2$. $5(3a - 2b)$

Solution: Substitute 1 for a, and 2 for b, then:

$$5(3a - 2b) = 15a - 10b = 15(1) - 10(2) = 15 - 20 = -5$$

4) *Solve this expression.* $4x - 3y,\ x = 3,\ y = 5$

Solution: Substitute 3 for x, and 5 for y and simplify. Then: $4x - 3y = 4(3) - 3(5) = 12 - 15 = -3$

Chapter 6: Practices

Simplify each expression.

1) $(6x - 4x + 8 + 6) =$

2) $(-14x + 26x - 12) =$

3) $(24x - 6 - 18x + 3) =$

4) $5 + 8x^2 - 9 =$

5) $7x - 4x^2 + 6x =$

6) $15x^2 - 3x - 6x^2 + 4 =$

Simplify each polynomial.

7) $2x^2 + 5x^3 - 7x^2 + 12x =$ ______________________

8) $2x^4 - 5x^5 + 8x^4 - 8x^2 =$ ______________________

9) $5x^3 + 15x - x^2 - 2x^3 =$ ______________________

10) $(8x^3 - 6x^2) + (9x^2 - 10x) =$ ______________________

11) $(12x^4 + 4x^3) - (8x^3 - 2x^4) =$ ______________________

12) $(9x^5 - 7x^3) - (5x^3 + x^2) =$ ______________________

Use the distributive property to simply each expression.

13) $4(5 + 6x) =$

14) $5(8 - 4x) =$

15) $(-6)(2 - 9x) =$

16) $(-7)(6x - 4) =$

17) $(3x + 12)4 =$

18) $(8x - 5)(-3) =$

Evaluate each expression using the value given.

19) $8 - x, x = -3$

20) $x + 12, x = -6$

21) $5x - 3, x = 2$

22) $4 - 6x, x = 1$

23) $3x + 1 , x = -2$

24) $15 - 2x, x = 5$

Evaluate each expression using the values given.

25) $4x - 2y,\ x = 4, y = -2$

26) $6a + 3b,\ a = 2, b = 4$

27) $12x - 5y - 8,\ x = 2, y = 3$

28) $-7a + 3b + 9,\ a = 4, b = 6$

29) $2x + 14 + 4y,\ x = 6, y = 8$

30) $4a - (5a - b) + 5, a = 4, b = 6$

Answers – Chapter 6

1) $2x + 14$
2) $12x - 12$
3) $6x - 3$
4) $8x^2 - 4$
5) $-4x^2 + 13x$
6) $9x^2 - 3x + 4$

7) $5x^3 - 5x^2 + 12x$
8) $-5x^5 + 10x^4 - 8x^2$
9) $3x^3 - x^2 + 15x$
10) $8x^3 + 3x^2 - 10x$
11) $14x^4 - 4x^3$
12) $9x^5 - 12x^3 - x^2$

13) $24x + 20$
14) $-20x + 40$
15) $54x - 12$
16) $-42x + 28$
17) $12x + 48$
18) $-24x + 15$
19) 11
20) 6
21) 7
22) -2
23) -5
24) 5
25) 20
26) 24
27) 1
28) -1
29) 58
30) 7

Chapter 7:

Equations and Inequalities

Math Topics that you'll learn in this Chapter:

- ✓ One–Step Equations
- ✓ Multi–Step Equations
- ✓ System of Equations
- ✓ Graphing Single–Variable Inequalities
- ✓ One–Step Inequalities
- ✓ Multi–Step Inequalities

One–Step Equations

- The values of two expressions on both sides of an equation are equal. Example: $ax = b$. In this equation, ax is equal to b.
- Solving an equation means finding the value of the variable.
- You only need to perform one Math operation in order to solve the one-step equations.
- To solve one-step equation, find the inverse (opposite) operation is being performed.
- The inverse operations are:

- Addition and subtraction
- Multiplication and division

Examples:

1) *Solve this equation for* x. $3x = 18, x = ?$

Solution: Here, the operation is multiplication (variable x is multiplied by 3) and its inverse operation is division. To solve this equation, divide both sides of equation by 3:

$3x = 18 \rightarrow \frac{3x}{3} = \frac{18}{3} \rightarrow x = 6$

2) *Solve this equation.* $x + 15 = 0$, $x = ?$

Solution: In this equation 15 is added to the variable x. The inverse operation of addition is subtraction. To solve this equation, subtract 15 from both sides of the equation:
$x + 15 - 15 = 0 - 15$. Then simplify: $x + 15 - 15 = 0 - 15 \rightarrow x = -15$

3) *Solve this equation for* x. $x - 23 = 0$

Solution: Here, the operation is subtraction and its inverse operation is addition. To solve this equation, add 23 to both sides of the equation: $x + 23 - 23 = 0 - 23 \rightarrow x = -23$

Multi–Step Equations

- To solve a multi-step equation, combine "like" terms on one side.
- Bring variables to one side by adding or subtracting.
- Simplify using the inverse of addition or subtraction.
- Simplify further by using the inverse of multiplication or division.
- Check your solution by plugging the value of the variable into the original equation.

Examples:

1) *Solve this equation for x.* $3x + 6 = 16 - 2x$

Solution: First bring variables to one side by adding $2x$ to both sides. Then:

$3x + 6 = 16 - 2x \rightarrow 3x + 6 + 2x = 16 - 2x + 2x$. Simplify: $5x + 6 = 16$

Now, subtract 6 from both sides of the equation: $5x + 6 - 6 = 16 - 6 \rightarrow 5x = 10 \rightarrow$

Divide both sides by 5: $5x = 10 \rightarrow \frac{5x}{5} = \frac{10}{5} \rightarrow x = 2$

Let's check this solution by substituting the value of 2 for x in the original equation:

$$x = 2 \rightarrow 3x + 6 = 16 - 2x \rightarrow 3(2) + 6 = 16 - 2(2) \rightarrow 6 + 6 = 16 - 4 \rightarrow 12 = 12$$

The answer $x = 2$ is correct.

2) *Solve this equation for x.* $-4x + 4 = 16$

Solution: Subtract 4 from both sides of the equation. $-4x + 4 - 4 = 16 - 4 \rightarrow -4x = 12$

Divide both sides by -4, then: $-4x = 12 \rightarrow \frac{-4x}{-4} = \frac{12}{-4} \rightarrow x = -3$

Now, check the solution: $x = -3 \rightarrow -4x + 4 = 16 \rightarrow -4(-3) + 4 = 16 \rightarrow 16 = 16$

The answer $x = -2$ is correct.

System of Equations

- A system of equations contains two equations and two variables. For example, consider the system of equations: $x - y = 1, x + y = 5$
- The easiest way to solve a system of equations is using the elimination method. The elimination method uses the addition property of equality. You can add the same value to each side of an equation.
- For the first equation above, you can add $x + y$ to the left side and 5 to the right side of the first equation: $x - y + (x + y) = 1 + 5$. Now, if you simplify, you get: $x - y + (x + y) = 1 + 5 \rightarrow 2x = 6 \rightarrow x = 3$. Now, substitute 3 for the x in the first equation: $3 - y = 1$. By solving this equation, $y = 2$

Example:

What is the value of $x + y$ *in this system of equations?* $\begin{cases} x + 2y = 6 \\ 2x - y = -8 \end{cases}$

Solution: Solving a System of Equations by Elimination:

Multiply the first equation by (-2), then add it to the second equation.

$$\frac{\begin{matrix} -2(x + 2y = 6) \\ 2x - y = -8 \end{matrix}}{} \Rightarrow \frac{\begin{matrix} -2x - 4y = -12 \\ 2x - y = -8 \end{matrix}}{} \Rightarrow -5y = -20 \Rightarrow y = 4$$

Plug in the value of y into one of the equations and solve for x.

$x + 2(4) = 6 \Rightarrow x + 8 = 6 \Rightarrow x = 6 - 8 \Rightarrow x = -2$

Thus, $x + y = -2 + 4 = 2$

Graphing Single–Variable Inequalities

- An inequality compares two expressions using an inequality sign.
- Inequality signs are: "less than" $<$, "greater than" $>$, "less than or equal to" $\leq$, and "greater than or equal to" $\geq$.
- To graph a single-variable inequality, find the value of the inequality on the number line.
- For less than ($<$) or greater than ($>$) draw open circle on the value of the variable. If there is an equal sign too, then use filled circle.
- Draw an arrow to the right for greater or to the left for less than.

Examples:

1) Draw a graph for this inequality. $x > 3$

Solution: Since, the variable is greater than 3, then we need to find 3 in the number line and draw an open circle on it.

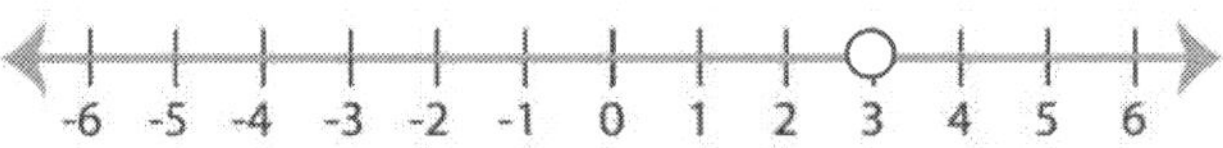

Then, draw an arrow to the right.

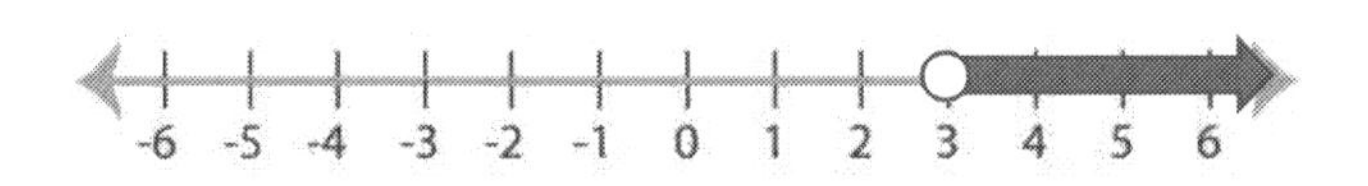

2) Graph this inequality. $x \leq -4$.

Solution: Since, the variable is less than or equal to -4, then we need to find -4 in the number line and draw a filled circle on it. Then, draw an arrow to the left.

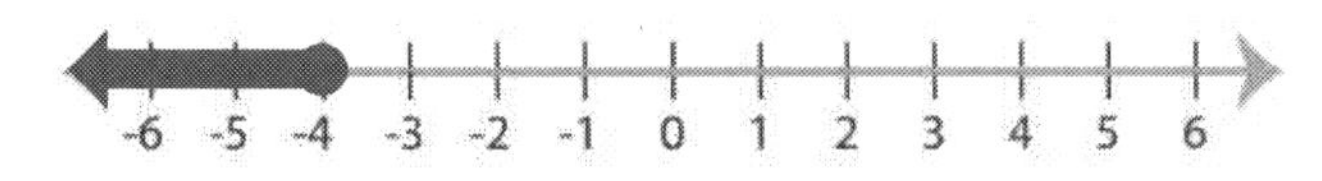

One–Step Inequalities

- An inequality compares two expressions using an inequality sign.
- Inequality signs are: "less than" $<$, "greater than" $>$, "less than or equal to" $\leq$, and "greater than or equal to" $\geq$.
- You only need to perform one Math operation in order to solve the one-step inequalities.
- To solve one-step inequalities, find the inverse (opposite) operation is being performed.
- For dividing or multiplying both sides by negative numbers, flip the direction of the inequality sign.

Examples:

1) *Solve this inequality for* x. $x + 3 \geq 4$

Solution: The inverse (opposite) operation of addition is subtraction. In this inequality, 3 is added to x. *To isolate* x *we need to* subtract 3 from both sides of the inequality. Then:

$x + 3 \geq 4 \rightarrow x + 3 - 3 \geq 4 - 3 \rightarrow x \geq 1$. The solution is: $x \geq 1$

2) *Solve the inequality.* $x - 5 > -4$.

Solution: 5 is subtracted from x. Add 5 to both sides. $x - 5 > -4 \rightarrow x - 5 + 5 > -4 + 5 \rightarrow x > 1$

3) *Solve.* $2x \leq -4$.

Solution: 2 is multiplied to x. Divide both sides by 2. Then: $2x \leq -4 \rightarrow \frac{2x}{2} \leq \frac{-4}{2} \rightarrow x \leq -2$

4) *Solve.* $-6x \leq 12$.

Solution: -6 is multiplied to x. Divide both sides by -6. Remember when dividing or multiplying both sides of an inequality by negative numbers, flip the direction of the inequality sign. Then:

$$-6x \leq 12 \rightarrow \frac{-6x}{-6} \geq \frac{12}{-6} \rightarrow x \geq -2$$

Multi–Step Inequalities

- To solve a multi-step inequality, combine "like" terms on one side.
- Bring variables to one side by adding or subtracting.
- Isolate the variable.
- Simplify using the inverse of addition or subtraction.
- Simplify further by using the inverse of multiplication or division.
- For dividing or multiplying both sides by negative numbers, flip the direction of the inequality sign.

1) *Solve this inequality.* $2x - 3 \leq 5$

Solution: In this inequality, 3 is subtracted from $2x$. The inverse of subtraction is addition. Add 3 to both sides of the inequality: $2x - 3 + 3 \leq 5 + 3 \rightarrow 2x \leq 8$

Now, divide both sides by 2. Then: $2x \leq 8 \rightarrow \frac{2x}{2} \leq \frac{8}{2} \rightarrow x \leq 4$

The solution of this inequality is $x \leq 4$.

2) *Solve this inequality.* $3x + 9 < 12$

Solution: First subtract 9 from both sides: $3x + 9 - 9 < 12 - 9$

Then simplify: $3x + 9 - 9 < 12 - 9 \rightarrow 3x < 3$

Now divide both sides by 3: $\frac{3x}{3} < \frac{3}{3} \rightarrow x < 1$

3) *Solve this inequality.* $-2x + 4 \geq 6$

First subtract 4 from both sides: $-2x + 4 - 4 \geq 6 - 4 \rightarrow -2x \geq 2$

Divide both sides by -2. Remember that you need to flip the direction of inequality sign.

$$-2x \geq 2 \rightarrow \frac{-2x}{-2} \leq \frac{2}{-2} \rightarrow x \leq -1$$

Chapter 7: Practices

Solve each equation. (One–Step Equations)

1) $x + 7 = 6, x =$ ____

2) $8 = 2 - x, x =$ ____

3) $-10 = 8 + x, x =$ ____

4) $x - 5 = -1, x =$ ____

5) $16 = x + 9, x =$ ____

6) $12 - x = -5, x =$ ____

Solve each equation. (Multi–Step Equations)

7) $5(x + 3) = 20$

8) $-4(7 - x) = 16$

9) $8 = -2\ (x + 5)$

10) $14 = 3(4 - 2x)$

11) $5(x + 7) = -10$

12) $-2(6 + 3x) = 12$

Solve each system of equations.

13) $-5x + y = -3$ $\quad x =$
$3x - 8y = 24$ $\quad y =$

14) $3x - 2y = 2$ $\quad x =$
$x - y = 2$ $\quad y =$

15) $4x + 7y = 2$ $\quad x =$
$6x + 7y = 10$ $\quad y =$

16) $5x + 7y = 18$ $\quad x =$
$-3x + 7y = -22$ $\quad y =$

Draw a graph for each inequality.

17) $x \leq -2$

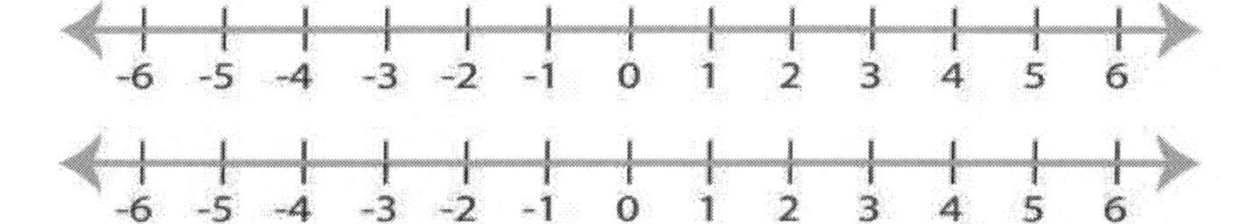

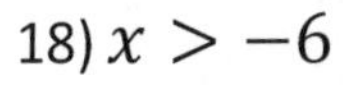
18) $x > -6$

Solve each inequality and graph it.

19) $x - 3 \geq -1$

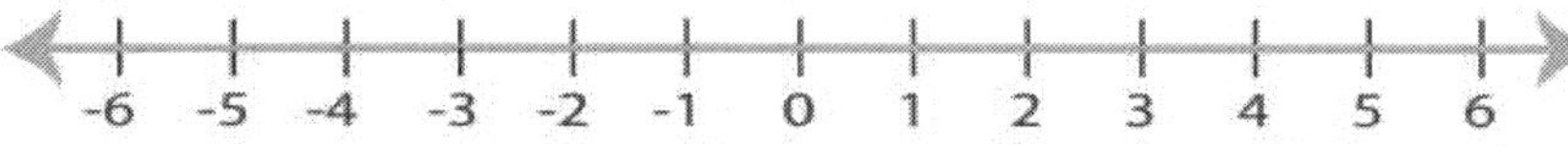

20) $3x - 2 < 16$

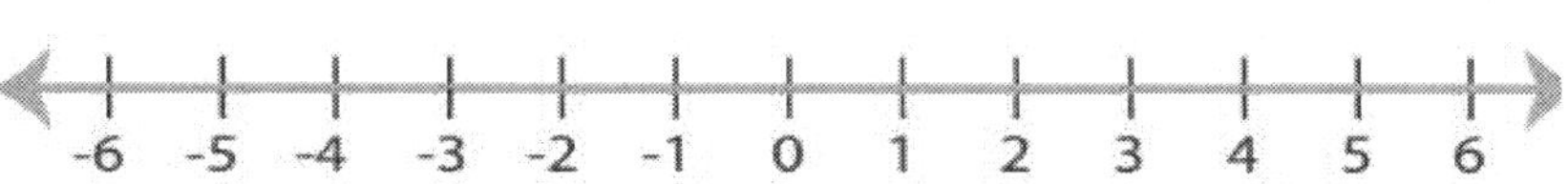

Solve each inequality.

21) $3x + 15 > -6$

22) $-18 + 4x \leq 10$

23) $4(x + 5) \geq 8$

24) $7x - 16 < 12$

25) $3(9 + x) \geq 15$

26) $-8 + 6x > 22$

Answers – Chapter 7

1) -1
2) -6
3) -18
4) 4
5) 7
6) 17

7) 1
8) 11
9) -9
10) $-\frac{1}{3}$
11) -9
12) -4

13) $x = 0\ , y = -3$

14) $x = -2, y = -4$

15) $x = 4, y = -2$

16) $x = 5, y = -1$

17) 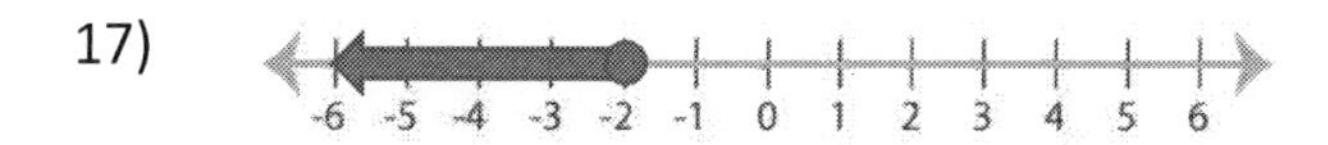

18)

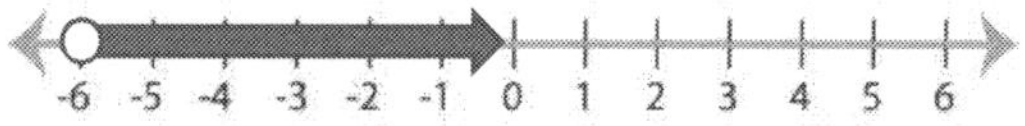

19)

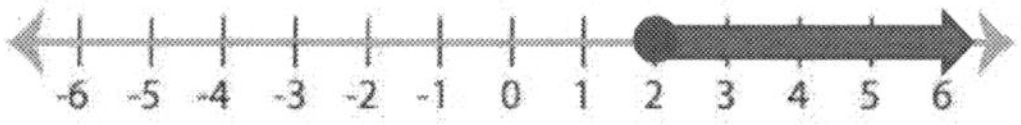

20)

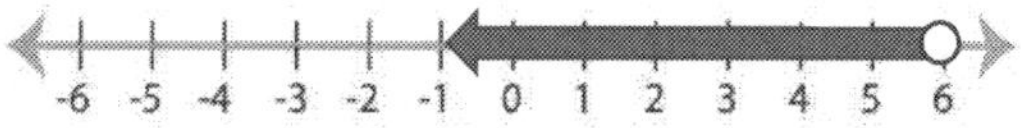

21) $x > -7$
22) $x \leq 7$
23) $x \geq -3$
24) $x < 4$
25) $x \geq -4$
26) $x > 5$

Chapter 8:

Lines and Slope

Math Topics that you'll learn in this Chapter:

- ✓ Finding Slope
- ✓ Graphing Lines Using Slope–Intercept Form
- ✓ Writing Linear Equations
- ✓ Graphing Linear Inequalities
- ✓ Finding Midpoint
- ✓ Finding Distance of Two Points

Finding Slope

- The slope of a line represents the direction of a line on the coordinate plane.
- A coordinate plane contains two perpendicular number lines. The horizontal line is x and the vertical line is y. The point at which the two axes intersect is called the origin. An ordered pair (x, y) shows the location of a point.
- A line on coordinate plane can be drawn by connecting two points.
- To find the slope of a line, we need the equation of the line or two points on the line.
- The slope of a line with two points A (x_1, y_1) and B (x_2, y_2) can be found by using this formula: $\frac{y_2 - y_1}{x_2 - x_1} = \frac{rise}{\text{run}}$
- The equation of a line is typically written as $y = mx + b$ where m is the slope and b is the y-intercept.

Examples:

1) *Find the slope of the line through these two points:* $A(2, -7)$ *and* $B(4, 3)$.

Solution: Slope $= \frac{y_2 - y_1}{x_2 - x_1}$. Let (x_1, y_1) be $A(2, -7)$ and (x_2, y_2) be $B(4, 3)$. (Remember, you can choose any point for (x_1, y_1) and (x_2, y_2)). *Then:* slope $= \frac{y_2 - y_1}{x_2 - x_1} = \frac{3-(-7)}{4-2} = \frac{10}{2} = 5$

The slope of the line through these two points is 5.

2) *Find the slope of the line with equation* $y = 3x + 6$

Solution: when the equation of a line is written in the form of $y = mx + b$, the slope is m. In this line: $y = 3x + 6$, the slope is 3.

Graphing Lines Using Slope–Intercept Form

- ✓ Slope-intercept form of a line: given the slope m and the y-intercept (the intersection of the line and y-axis) b, then the equation of the line is: $y = mx + b$
- ✓ To draw the graph of a linear equation in slope-intercept form on the xy coordinate plane, find two points on the line by plugging two values for x and calculating the values of y.
- ✓ You can also use the slope (m) and one point to graph the line.

Example:

1) *Sketch the graph of* $y = 2x - 4$.

Solution: To graph this line, we need to find two points. When x is zero the value of y is -4. And when x is 2 the value of y is 0.

$$x = 0 \rightarrow y = 2(0) - 4 = -4,$$

$$y = 0 \rightarrow 0 = 2x - 4 \rightarrow x = 2$$

Now, we have two points: $(0, -4)$ and $(2, 0)$.

Find the points on the coordinate plane and graph the line. Remember that the slope of the line is 2.

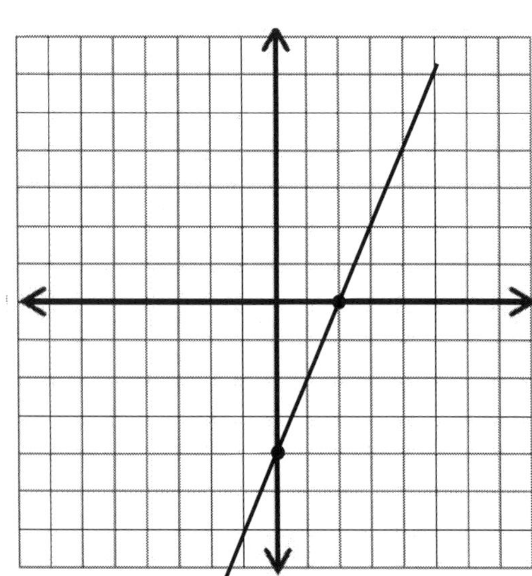

Writing Linear Equations

- The equation of a line in slope-intercept form: $\boldsymbol{y = mx + b}$
- To write the equation of a line, first identify the slope.
- Find the y-intercept. This can be done by substituting the slope and the coordinates of a point $(\boldsymbol{x}, \boldsymbol{y})$ on the line.

1) What is the equation of the line that passes through $(\mathbf{2}, -\mathbf{4})$ and has a slope of $\mathbf{8}$?

Solution: The general slope-intercept form of the equation of a line is $y = mx + b$, where m is the slope and b is the y-intercept.
By substitution of the given point and given slope: $y = mx + b \rightarrow -4 = (2)(8) + b$
So, $b = -4 - 16 = -20$, and the required equation is $y = 8x - 20$

2) Write the equation of the line through two points $A(2, 1)$ and $B(-2, 5)$.

Solution: Fist find the slope: $Slop = \frac{y_2 - y_1}{x_2 - x_1} = \frac{5 - 1}{-2 - 2} = \frac{4}{-4} = -1 \rightarrow m = -1$
To find the value of b, use either points and plug in the values of x and y in the equation. The answer will be the same: $y = -x + b$. Let's check both points. Then:
$(2, 1) \rightarrow y = mx + b \rightarrow 1 = -1(2) + b \rightarrow b = 3$
$(-2,5) \rightarrow y = mx + b \rightarrow 5 = -1(-2) + b \rightarrow b = 3$. The y-intercept of the line is 3.
The equation of the line is: $y = -x + 3$

3) What is the equation of the line that passes through $(\mathbf{2}, -\mathbf{1})$ and has a slope of $\mathbf{5}$?

Solution: The general slope-intercept form of the equation of a line is $y = mx + b$, where m is the slope and b is the y-intercept.
By substitution of the given point and given slope: $y = mx + b \rightarrow -1 = (5)(2) + b$
So, $b = -1 - 10 = -11$, and the equation of the line is: $y = 5x - 11$.

Finding Midpoint

- The middle of a line segment is its midpoint.
- The Midpoint of two endpoints A (x_1, y_1) and B (x_2, y_2) can be found using this formula: M $(\frac{x_1+x_2}{2}, \frac{y_1+y_2}{2})$

1) Find the midpoint of the line segment with the given endpoints. $(\mathbf{1}, -\mathbf{3}), (\mathbf{3}, \mathbf{7})$

Solution: Midpoint $= (\frac{x_1+x_2}{2}, \frac{y_1+y_2}{2}) \rightarrow (x_1, y_1) = (1, -3)$ and $(x_2, y_2) = (3, 7)$

Midpoint $= (\frac{1+3}{2}, \frac{-3+7}{2}) \rightarrow (\frac{4}{2}, \frac{4}{2}) \rightarrow M(2, 2)$

2) Find the midpoint of the line segment with the given endpoints. $(-\mathbf{4}, \mathbf{5}), (\mathbf{8}, -\mathbf{7})$

Solution: Midpoint $= (\frac{x_1+x_2}{2}, \frac{y_1+y_2}{2}) \rightarrow (x_1, y_1) = (-4, 5)$ and $(x_2, y_2) = (8, -7)$

Midpoint $= (\frac{-4+8}{2}, \frac{5-7}{2}) \rightarrow (\frac{4}{2}, \frac{-2}{2}) \rightarrow M(2, -1)$

3) Find the midpoint of the line segment with the given endpoints. $(\mathbf{5}, -\mathbf{2}), (\mathbf{1}, \mathbf{10})$

Solution: Midpoint $= (\frac{x_1+x_2}{2}, \frac{y_1+y_2}{2}) \rightarrow (x_1, y_1) = (5, -2)$ and $(x_2, y_2) = (1,10)$

Midpoint $= (\frac{5+1}{2}, \frac{-2+10}{2}) \rightarrow (\frac{6}{2}, \frac{8}{2}) \rightarrow M(3, 4)$

4) Find the midpoint of the line segment with the given endpoints. $(\mathbf{2}, \mathbf{3}), (\mathbf{12}, -\mathbf{9})$

Solution: Midpoint $= (\frac{x_1+x_2}{2}, \frac{y_1+y_2}{2}) \rightarrow (x_1, y_1) = (2, 3)$ and $(x_2, y_2) = (12, -3)$

Midpoint $= (\frac{2+12}{2}, \frac{3-9}{2}) \rightarrow (\frac{14}{2}, \frac{-6}{2}) \rightarrow M(7, -3)$

Finding Distance of Two Points

✓ Use following formula to find the distance of two points with the coordinates A (x_1, y_1) and B (x_2, y_2):

$$d = \sqrt{(x_2 - x_1)^2 + (y_2 - y_1)^2}$$

Examples:

1) Find the distance between $(4, 6)$ *and* $(1, 2)$.

Solution: *Use distance of two points formula:* $d = \sqrt{(x_2 - x_1)^2 + (y_2 - y_1)^2}$

$(x_1, y_1) = (4, 6)$ and $(x_2, y_2) = (1, 2)$. *Then:* $d = \sqrt{(x_2 - x_1)^2 + (y_2 - y_1)^2} \rightarrow$

$$d = \sqrt{(1-(4))^2 + (2-6)^2} = \sqrt{(-3)^2 + (-4)^2} = \sqrt{9+16} = \sqrt{25} = 5 \rightarrow d = 5$$

2) Find the distance of two points $(-6, -10)$ *and* $(-2, -10)$.

Solution: *Use distance of two points formula:* $d = \sqrt{(x_2 - x_1)^2 + (y_2 - y_1)^2}$

$(x_1, y_1) = (-6, -10)$, and $(x_2, y_2) = (-2, -10)$

Then: $d = \sqrt{(x_2 - x_1)^2 + (y_2 - y_1)^2} \rightarrow d = \sqrt{(-2-(-6))^2 + (-10-(-10))^2} =$

$\sqrt{(4)^2 + (0)^2} = \sqrt{16+0} = \sqrt{16} = 4$. Then: $d = 4$

3) Find the distance between $(-6, 5)$ *and* $(-2, 2)$.

Solution: *Use distance of two points formula:* $d = \sqrt{(x_2 - x_1)^2 + (y_2 - y_1)^2}$

$(x_1, y_1) = (-6, 5)$ and $(x_2, y_2) = (-2, 2)$. *Then:* $d = \sqrt{(x_2 - x_1)^2 + (y_2 - y_1)^2} \rightarrow$

$$d = \sqrt{(-2-(-6))^2 + (2-5)^2} = \sqrt{(4)^2 + (-3)^2} = \sqrt{16+9} = \sqrt{25} = 5$$

Chapter 8: Practices

Find the slope of each line.

1) $y = x - 1$
2) $y = -2x + 5$
3) $y = 2x - 1$
4) *Line through* $(-2, 4)$ *and* $(6, 0)$
5) *Line through* $(-3, 5)$ *and* $(-2, 8)$
6) *Line through* $(-3, -1)$ *and* $(0, -4)$

Sketch the graph of each line. (Using Slope–Intercept Form)

7) $y = x + 2$

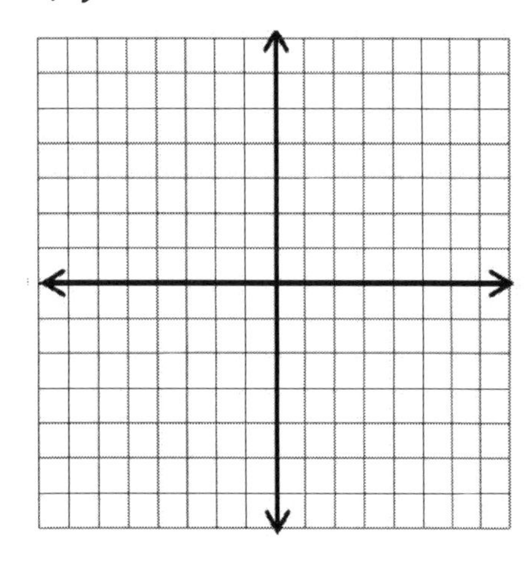

8) $y = 2x - 3$

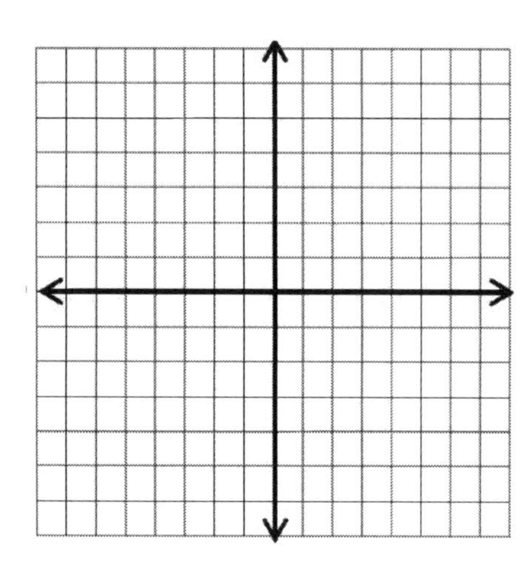

Solve.

9) What is the equation of a line with slope 4 and intercept 12? ________________

10) What is the equation of a line with slope 4 and passes through point $(4, 2)$?

____

11) What is the equation of a line with slope -2 and passes through point $(-2, 4)$?

____

12) The slope of a line is -3 and it passes through point $(-1, 5)$. What is the equation of the line? ________________

13) The slope of a line is 3 and it passes through point $(-1, 4)$. What is the equation of the line? ________________

Sketch the graph of each linear inequality. (Graphing Linear Inequalities)

15) $y > 3x - 1$

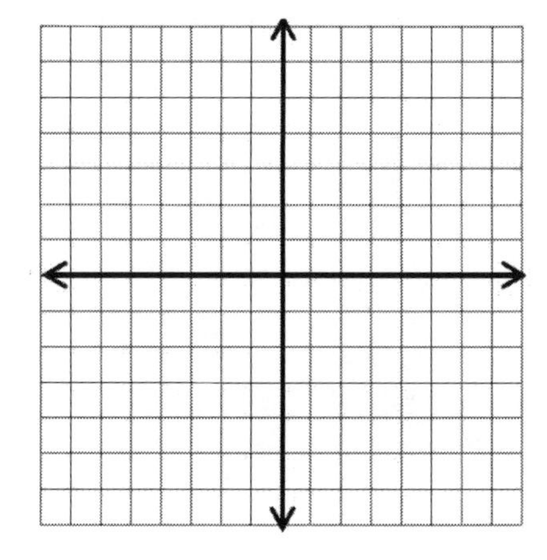

16) $y < -x + 4$

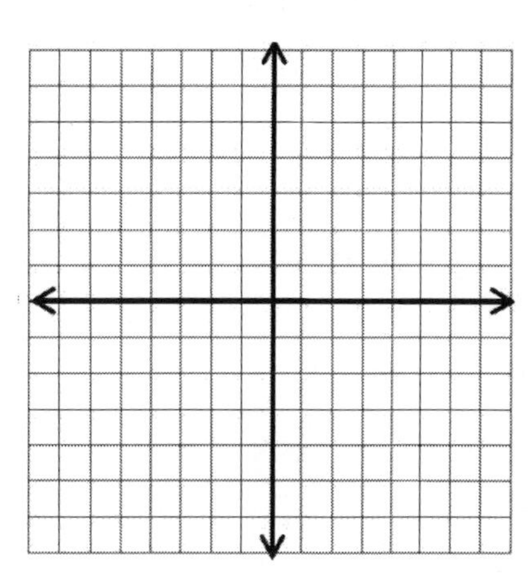

Find the midpoint of the line segment with the given endpoints.

17) $(2, 5), (-2, 7)$
18) $(-3, 6), (5, 2)$
19) $(9, -7), (-6, 5)$
20) $(14, -8), (-4, -2)$
21) $(-11, 2), (3, 8)$
22) $(7, 10), (1, -6)$

Find the distance between each pair of points.

23) $(3, 10), (-2, -2)$
24) $(1, 6), (4, 10)$
25) $(-2, -1), (-8, 7)$
26) $(8, -2), (5, -6)$
27) $(4, -3), (-5, 9)$
28) $(0, 6), (3, 2)$

Answers – Chapter 8

Find the slope of the line through each pair of points.

1) 1
2) -2
3) 2
4) $-\frac{1}{2}$
5) 3
6) -1

Sketch the graph of each line. (Using Slope–Intercept Form)

7) $y = x + 2$

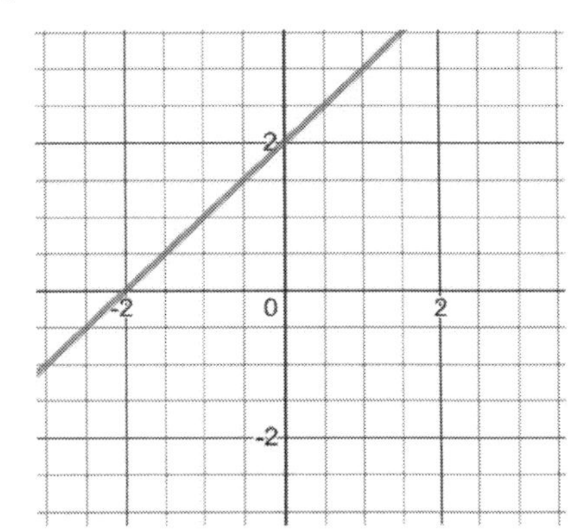

8) $y = 2x - 3$

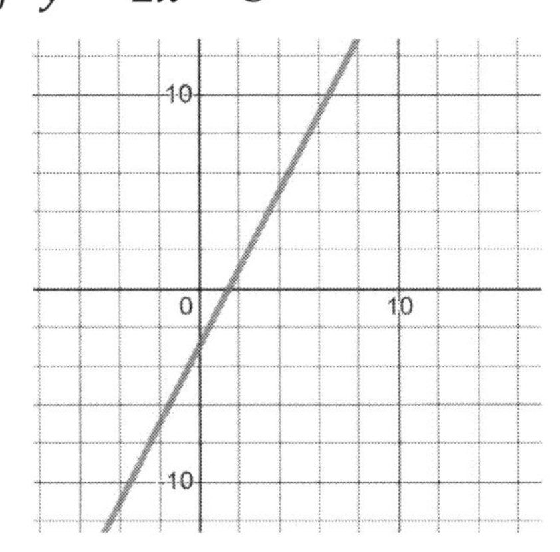

Write the equation of the line through the given points.

9) $y = 4x + 12$
10) $y = 4x - 14$
11) $y = -2x + 8$
12) $y = -3x + 2$
13) $y = 3x + 7$

Sketch the graph of each linear inequality. (Graphing Linear Inequalities)

15) $y > 3x - 1$

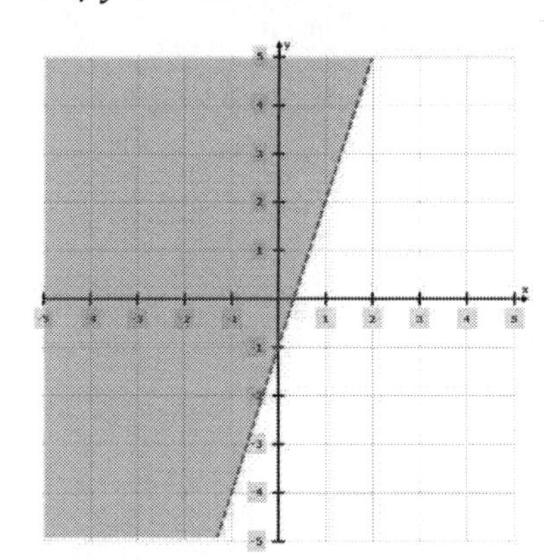

16) $y < -x + 4$

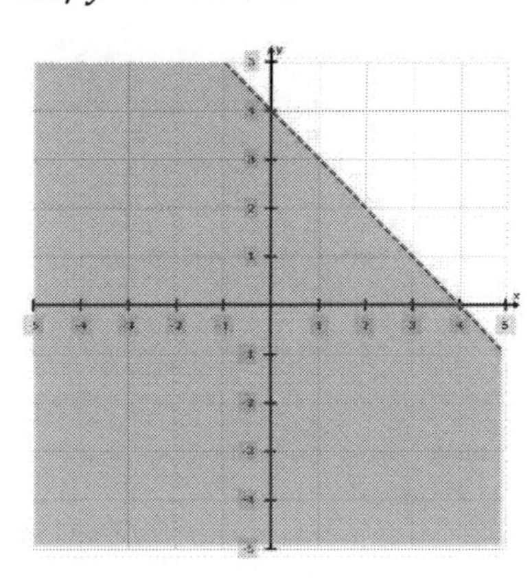

Find the midpoint of the line segment with the given endpoints.

17) $(0, 6)$

18) $(1, 4)$

19) $(1.5, -1)$

20) $(5, -5)$

21) $(-4, 5)$

22) $(4, 2)$

Find the distance between each pair of points.

23) 13

24) 5

25) 10

26) 5

27) 15

28) 5

Chapter 9:

Exponents and Variables

Math Topics that you'll learn in this Chapter:

- ✓ Multiplication Property of Exponents
- ✓ Division Property of Exponents
- ✓ Powers of Products and Quotients
- ✓ Zero and Negative Exponents
- ✓ Negative Exponents and Negative Bases
- ✓ Scientific Notation
- ✓ Radicals

Multiplication Property of Exponents

- Exponents are shorthand for repeated multiplication of the same number by itself. For example, instead of 2×2, we can write 2^2. For $3 \times 3 \times 3 \times 3$, we can write 3^4
- In algebra, a variable is a letter used to stand for a number. The most common letters are: $x, y, z, a, b, c, m,$ and n.
- Exponent's rules: $x^a \times x^b = x^{a+b}$, $\frac{x^a}{x^b} = x^{a-b}$

$(x^a)^b = x^{a \times b}$ $\quad$ $(xy)^a = x^a \times y^a$ $\quad$ $(\frac{a}{b})^c = \frac{a^c}{b^c}$

Examples:

1) *Multiply.* $4x^3 \times 2x^2$

Solution: Use Exponent's rules: $x^a \times x^b = x^{a+b} \rightarrow x^3 \times x^2 = x^{3+2} = x^5$

Then: $4x^3 \times 2x^2 = 8x^5$

2) *Simplify.* $(x^3y^5)^2$

Solution: Use Exponent's rules: $(x^a)^b = x^{a \times b}$. Then: $(x^3y^5)^2 = x^{3\times2}y^{5\times2} = x^6y^{10}$

3) *Multiply.* $-2x^5 \times 7x^3$

Solution: Use Exponent's rules: $x^a \times x^b = x^{a+b} \rightarrow x^5 \times x^3 = x^{5+3} = x^8$

Then: $-2x^5 \times 7x^3 = -14x^8$

4) *Simplify.* $(x^2y^4)^3$

Solution: Use Exponent's rules: $(x^a)^b = x^{a \times b}$. Then: $(x^2y^4)^3 = x^{2\times3}y^{4\times3} = x^6y^{12}$

Division Property of Exponents

- Exponents are shorthand for repeated multiplication of the same number by itself. For example, instead of 3×3, we can write 3^2. For $2 \times 2 \times 2$, we can write 2^3
- For division of exponents use following formulas:

$$\frac{x^a}{x^b} = x^{a-b}, x \neq 0, \quad \frac{x^a}{x^b} = \frac{1}{x^{b-a}}, x \neq 0, \quad \frac{1}{x^b} = x^{-b}$$

Examples:

1) *Simplify.* $\frac{12x^2y}{4xy^3} =$

Solution: First cancel the common factor: $4 \rightarrow \frac{12x^2y}{4xy^3} = \frac{3x^2y}{xy^3}$

Use Exponent's rules: $\frac{x^a}{x^b} = x^{a-b} \rightarrow \frac{x^2}{x} = x^{2-1} = x$ and $\frac{y}{y^3} = \frac{1}{y^{3-1}} = \frac{1}{y^2}$

Then: $\frac{12x^2y}{4xy^3} = \frac{3x}{y^2}$

2) *Simplify.* $\frac{18x^6}{2x^3} =$

Solution: Use Exponent's rules: $\frac{x^a}{x^b} = x^{b-a} \rightarrow \frac{x^6}{x^3} = x^{6-3} = x^3$

Then: $\frac{18x^6}{2x^3} = 9x^3$

3) *Simplify.* $\frac{8x^3y}{40x^2y^3} =$

Solution: First cancel the common factor: $8 \rightarrow \frac{8x^3y}{40x^2y^3} = \frac{x^3y}{5x^2y^3}$

Use Exponent's rules: $\frac{x^a}{x^b} = x^{a-b} \rightarrow \frac{x^3}{x^2} = x^{3-2} = x$

Then: $\frac{8x^3y}{40x^2y^3} = \frac{xy}{5y^3} \rightarrow$ now cancel the common factor: $y \rightarrow \frac{xy}{5y^3} = \frac{x}{5y^2}$

Powers of Products and Quotients

- Exponents are shorthand for repeated multiplication of the same number by itself. For example, instead of $2 \times 2 \times 2$, we can write 2^3. For $3 \times 3 \times 3 \times 3$, we can write 3^4
- For any nonzero numbers a and b and any integer x, $(ab)^x = a^x \times b^x$ and $(\frac{a}{b})^c = \frac{a^c}{b^c}$

Examples:

1) *Simplify.* $(6x^2y^4)^2$

Solution: Use Exponent's rules: $(x^a)^b = x^{a \times b}$

$(6x^2y^4)^2 = (6)^2(x^2)^2(y^4)^2 = 36x^{2\times2}y^{4\times2} = 36x^4y^8$

2) *Simplify.* $(\frac{5x}{2x^2})^2$

Solution: First cancel the common factor: $x \rightarrow \left(\frac{5x}{2x^2}\right)^2 = \left(\frac{5}{2x}\right)^2$

Use Exponent's rules: $(\frac{a}{b})^c = \frac{a^c}{b^c}$, Then: $(\frac{5}{2x})^2 = \frac{5^2}{(2x)^2} = \frac{25}{4x^2}$

3) *Simplify.* $(3x^5y^4)^2$

Solution: Use Exponent's rules: $(x^a)^b = x^{a \times b}$

$(3x^5y^4)^2 = (3)^2(x^5)^2(y^4)^2 = 9x^{5\times2}y^{4\times2} = 9x^{10}y^8$

4) *Simplify.* $(\frac{2x}{3x^2})^2$

Solution: First cancel the common factor: $x \rightarrow \left(\frac{2x}{3x^2}\right)^2 = \left(\frac{2}{3x}\right)^2$

Use Exponent's rules: $(\frac{a}{b})^c = \frac{a^c}{b^c}$, Then: $(\frac{2}{3x})^2 = \frac{2^2}{(3x)^2} = \frac{4}{9x^2}$

Zero and Negative Exponents

- Zero-Exponent Rule: $a^0 = 1$, this means that anything raised to the zero power is 1. For example: $(5xy)^0 = 1$
- A negative exponent simply means that the base is on the wrong side of the fraction line, so you need to flip the base to the other side. For instance, "x^{-2}" (pronounced as "ecks to the minus two") just means "x^2" but underneath, as in $\frac{1}{x^2}$.

Examples:

1) *Evaluate.* $(\frac{2}{3})^{-2} =$

Solution: Use negative exponent's rule: $\left(\frac{x^a}{x^b}\right)^{-2} = \left(\frac{x^b}{x^a}\right)^2 \rightarrow \left(\frac{2}{3}\right)^{-2} = \left(\frac{3}{2}\right)^2 =$

Then: $\left(\frac{3}{2}\right)^2 = \frac{3^2}{2^2} = \frac{9}{4}$

2) *Evaluate.* $(\frac{4}{5})^{-3} =$

Solution: Use negative exponent's rule: $\left(\frac{x^a}{x^b}\right)^{-2} = \left(\frac{x^b}{x^a}\right)^2 \rightarrow \left(\frac{4}{5}\right)^{-3} = \left(\frac{5}{4}\right)^3 =$

Then: $\left(\frac{5}{4}\right)^3 = \frac{5^3}{4^3} = \frac{125}{64}$

3) *Evaluate.* $(\frac{x}{y})^0 =$

Solution: Use zero-exponent Rule: $a^0 = 1$

Then: $(\frac{x}{y})^0 = 1$

4) *Evaluate.* $(\frac{5}{6})^{-1} =$

Solution: Use negative exponent's rule: $\left(\frac{x^a}{x^b}\right)^{-2} = \left(\frac{x^b}{x^a}\right)^2 \rightarrow \left(\frac{5}{6}\right)^{-1} = \left(\frac{6}{5}\right)^1 = \frac{6}{5}$

Negative Exponents and Negative Bases

- A negative exponent is the reciprocal of that number with a positive exponent. $(3)^{-2} = \frac{1}{3^2}$
- To simplify a negative exponent, make the power positive!
- The parenthesis is important! -5^{-2} is not the same as $(-5)^{-2}$

$$-5^{-2} = -\frac{1}{5^2} \text{ and } (-5)^{-2} = +\frac{1}{5^2}$$

Examples:

1) *Simplify.* $(\frac{5a}{6c})^{-2} =$

Solution: Use negative exponent's rule: $(\frac{x^a}{x^b})^{-2} = (\frac{x^b}{x^a})^2 \rightarrow (\frac{5a}{6c})^{-2} = (\frac{6c}{5a})^2$

Now use exponent's rule: $(\frac{a}{b})^c = \frac{a^c}{b^c} \rightarrow = (\frac{6c}{5a})^2 = \frac{6^2c^2}{5^2a^2}$

Then: $\frac{6^2c^2}{5^2a^2} = \frac{36c^2}{25a^2}$

2) *Simplify.* $(\frac{2x}{3yz})^{-3} =$

Solution: Use negative exponent's rule: $(\frac{x^a}{x^b})^{-2} = (\frac{x^b}{x^a})^2 \rightarrow (\frac{2x}{3yz})^{-3} = (\frac{3yz}{2x})^3$

Now use exponent's rule: $(\frac{a}{b})^c = \frac{a^c}{b^c} \rightarrow (\frac{3yz}{2x})^3 = \frac{3^3y^3z^3}{2^3x^3} = \frac{27y^3z^3}{8x^3}$

3) *Simplify.* $(\frac{3a}{2c})^{-2} =$

Solution: Use negative exponent's rule: $(\frac{x^a}{x^b})^{-2} = (\frac{x^b}{x^a})^2 \rightarrow (\frac{3a}{2c})^{-2} = (\frac{2c}{3a})^2$

Now use exponent's rule: $(\frac{a}{b})^c = \frac{a^c}{b^c} \rightarrow = (\frac{2c}{3a})^2 = \frac{2^2c^2}{3^2a^2}$

Then: $\frac{2^2c^2}{3^2a^2} = \frac{4c^2}{9a^2}$

Scientific Notation

- Scientific notation is used to write very big or very small numbers in decimal form.
- In scientific notation all numbers are written in the form of: $m \times 10^n$, where m is greater than 1 and less than 10.
- To convert a number from scientific notation to standard form, move the decimal point to the left (if the exponent of ten is a negative number), or to the right (if the exponent is positive).

Examples:

1) *Write* 0.00015 *in scientific notation.*

Solution: First, move the decimal point to the right so that you have a number that is between 1 and 10. That number is 1.5

Now, determine how many places the decimal moved in step 1 by the power of 10. We moved the decimal point 4 digits to the right. Then: 10^{-4} → When the decimal moved to the right, the exponent is negative. Then: $0.00015 = 1.5 \times 10^{-4}$

2) *Write* $\mathbf{9.5 \times 10^{-5}}$ *in standard notation.*

Solution: 10^{-5} → When the decimal moved to the right, the exponent is negative.

Then: $9.5 \times 10^{-5} = 0.000095$

3) *Write* $\mathbf{0.00012}$ *in scientific notation.*

Solution: First, move the decimal point to the right so that you have a number that is between 1 and 10. Then: $m = 1.2$

Now, determine how many places the decimal moved in step 1 by the power of 10.

10^{-4} → Then: $0.00012 = 1.2 \times 10^{-4}$

4) *Write* $\mathbf{8.3 \times 10^{5}}$ *in standard notation.*

Solution: 10^{-5} → The exponent is positive 5. Then, move the decimal point to the right five digits. (remember $8.3 = 8.30000$)

Then: $8.3 \times 10^{5} = 830000$

Radicals

- If n is a positive integer and x is a real number, then: $\sqrt[n]{x} = x^{\frac{1}{n}}$, $\sqrt[n]{xy} = x^{\frac{1}{n}} \times y^{\frac{1}{n}}$, $\sqrt[n]{\frac{x}{y}} = \frac{x^{\frac{1}{n}}}{y^{\frac{1}{n}}}$, and $\sqrt[n]{x} \times \sqrt[n]{y} = \sqrt[n]{xy}$
- A square root of x is a number r whose square is: $\boldsymbol{r^2 = x}$ ($\boldsymbol{r}$ is a square root of $\boldsymbol{x}$.
- To add and subtract radicals, we need to have the same values under the radical. For example: $\sqrt{3} + \sqrt{3} = 2\sqrt{3}$, $3\sqrt{5} - \sqrt{5} = 2\sqrt{5}$

Examples:

1) *Find the square root of* $\sqrt{169}$.

Solution: First factor the number: $169 = 13^2$,

Then: $\sqrt{169} = \sqrt{13^2}$

Now use radical rule: $\sqrt[n]{a^n} = a$.

Then: $\sqrt{169} = \sqrt{13^2} = 13$

2) *Evaluate.* $\sqrt{9} \times \sqrt{25} =$

Solution: Find the values of $\sqrt{9}$ and $\sqrt{25}$.

Then: $\sqrt{9} \times \sqrt{25} = 3 \times 5 = 15$

3) *Solve.* $7\sqrt{2} + 4\sqrt{2}$.

Solution: Since we have the same values under the radical, we can add these two radicals: $7\sqrt{2} + 4\sqrt{2} = 11\sqrt{2}$

4) *Evaluate.* $\sqrt{2} \times \sqrt{8} =$

Solution: Use this radical rule: $\sqrt[n]{x} \times \sqrt[n]{y} = \sqrt[n]{xy} \rightarrow \sqrt{2} \times \sqrt{8} = \sqrt{16}$

The square root of 16 is 4. Then: $\sqrt{2} \times \sqrt{8} = \sqrt{16} = 4$

Chapter 9: Practices

✍ ***Simplify and write the answer in exponential form.***

1) $3x^3 \times 5xy^2 =$
2) $4x^2y \times 6x^2y^2 =$
3) $8x^3y^2 \times 2x^2y^3 =$
4) $7xy^4 \times 3x^2y =$
5) $6x^4y^5 \times 8x^3y^2 =$
6) $5x^3y^3 \times 8x^3y^3 =$

✍ ***Simplify. (Division Property of Exponents)***

7) $\frac{5^5 \times 5^3}{5^9 \times 5} =$
8) $\frac{8x}{24x^2} =$
9) $\frac{15 \quad ^4}{9x^3} =$
10) $\frac{36 \quad ^3}{54x^3y^2} =$
11) $\frac{14y^3}{49x^4y^4} =$
12) $\frac{120x^3y^5}{30x^2y^3} =$

✍ ***Simplify. (Powers of Products and Quotients)***

13) $(8x^4y^6)^3 =$
14) $(3x^5y^4)^6 =$
15) $(5x \times 4xy^2)^2 =$
16) $(\frac{6x}{x^3})^2 =$
17) $\left(\frac{2x^3y^5}{6x^4y^2}\right)^2 =$
18) $\left(\frac{42x^4y^6}{21 \quad ^3y^5}\right)^3 =$

✍ ***Evaluate the following expressions. (Zero and Negative Exponents)***

19) $(\frac{2}{5})^{-2} =$
20) $(\frac{1}{2})^{-8} =$
21) $(\frac{2}{5})^{-3} =$
22) $(\frac{3}{7})^{-2} =$
23) $(\frac{5}{6})^{-3} =$
24) $(\frac{4}{9})^{-2} =$

✍ ***Simplify. (Negative Exponents and Negative Bases)***

25) $16x^{-3}y^{-4} =$
26) $-9x^2y^{-3} =$
27) $12a^{-4}b^2 =$
28) $25a^3b^{-5}c^{-1} =$
29) $\frac{18}{x^2y^{-2}} =$
30) $\frac{21 \quad ^{-2}b}{-14c^{-4}} =$

Write each number in scientific notation.

31) $0.00615 =$

32) $0.000048 =$

33) $36{,}000 =$

34) $82{,}000{,}000 =$

Evaluate.

35) $\sqrt{7} \times \sqrt{7} =$ ________

36) $\sqrt{36} - \sqrt{9} =$ ________

37) $\sqrt{25} + \sqrt{49} =$ ________

38) $\sqrt{16} \times \sqrt{64} =$ ________

39) $\sqrt{3} \times \sqrt{12} =$ ________

40) $2\sqrt{6} + 3\sqrt{6} =$ ________

Answers – Chapter 9

1) $15x^4y^2$
2) $24x^4y^3$
3) $16x^5y^5$
4) $21x^3y^5$
5) $48x^7y^7$
6) $40x^6y^6$
7) $\frac{1}{25}$
8) $\frac{1}{3x}$
9) $\frac{5}{3}x$
10) $\frac{2}{3y^3}$
11) $\frac{2}{7x^4y}$
12) $4xy^2$
13) $512x^{12}y^{18}$
14) $729x^{30}y^{24}$
15) $400x^4y^4$
16) $\frac{36}{x^4}$
17) $\frac{y^6}{9x^2}$
18) $8x^3y^3$
19) $\frac{25}{4}$
20) 256
21) $\frac{125}{8}$
22) $\frac{49}{9}$
23) $\frac{216}{125}$
24) $\frac{81}{16}$
25) $\frac{16}{x^3y^4}$
26) $-\frac{9x^2}{y^3}$
27) $\frac{12b^2}{a^4}$
28) $\frac{25\ \ ^3}{b^5c}$
29) $\frac{18y^3}{x^2}$
30) $-\frac{3bc^4}{2a^2}$
31) 6.15×10^{-3}
32) 4.8×10^{-5}
33) 3.6×10^4
34) 8.2×10^7
35) 7
36) 3
37) 12
38) 32
39) 6
40) $5\sqrt{6}$

Chapter 10: Polynomials

Math Topics that you'll learn in this Chapter:

- ✓ Simplifying Polynomials
- ✓ Adding and Subtracting Polynomials
- ✓ Multiplying Monomials
- ✓ Multiplying and Dividing Monomials
- ✓ Multiplying a Polynomial and a Monomial
- ✓ Multiplying Binomials
- ✓ Factoring Trinomials

Simplifying Polynomials

- ✓ To simplify Polynomials, find "like" terms. (they have same variables with same power).
- ✓ Use "FOIL". (First-Out-In-Last) for binomials:

$$(x + a)(x + b) = x^2 + (b + a)x + ab$$

- ✓ Add or Subtract "like" terms using order of operation.

Examples:

1) Simplify this expression. $x(2x + 5) + 6x =$

Solution: Use Distributive Property: $x(2x + 5) = 2x^2 + 5x$

Now, combine like terms: $x(2x + 5) + 6x = 2x^2 + 5x + 6x = 2x^2 + 11x$

2) Simplify this expression. $(x + 2)(x + 3) =$

Solution: First apply FOIL method: $(a + b)(c + d) = ac + ad + bc + bd$

$(x + 2)(x + 3) = x^2 + 3x + 2x + 6$

Now combine like terms: $x^2 + 3x + 2x + 6 \; = x^2 + 5x + 6$

3) Simplify this expression. $4x(2x - 3) + 6x^2 - 4x =$

Solution: Use Distributive Property: $4x(2x - 3) = 8x^2 - 12x$

Then: $4x(2x - 3) + 6x^2 - 4x = 8x^2 - 12x + 6x^2 - 4x$

Now combine like terms: $8x^2 + 6x^2 = 14x^2$, and $-12x - 4x = -16x$

The simplified form of the expression: $8x^2 - 12x + 6x^2 - 4x = 14x^2 - 16x$

Adding and Subtracting Polynomials

- Adding polynomials is just a matter of combining like terms, with some order of operations considerations thrown in.
- Be careful with the minus signs, and don't confuse addition and multiplication!
- For subtracting polynomials, sometimes you need to use the Distributive Property: $a(b+c) = ab + ac$, $a(b-c) = ab - ac$

Examples:

1) *Simplify the expressions.* $(x^3 - 3x^4) - (2x^4 - 5x^3) =$

Solution: First use Distributive Property: $-(2x^4 - 5x^3) = -1(2x^4 - 5x^3) = -2x^4 + 5x^3$

$\rightarrow (x^3 - 3x^4) - (2x^4 - 5x^3) = x^3 - 3x^4 - 2x^4 + 5x^3$

Now combine like terms: $x^3 + 5x^3 = 6x^3$ and $-3x^4 - 2x^4 = -5x^4$

Then: $(x^3 - 3x^4) - (2x^4 - 5x^3) = x^3 - 3x^4 - 2x^4 + 5x^3 = 6x^3 - 5x^4$

Write the answer in standard form: $6x^3 - 5x^4 = -5x^4 + 6x^3$

2) *Add expressions.* $(2x^3 - 4) + (6x^3 - 2x^2) =$

Solution: Remove parentheses: $(2x^3 - 4) + (6x^3 - 2x^2) = 2x^3 - 4 + 6x^3 - 2x^2$

Now combine like terms: $2x^3 - 4 + 6x^3 - 2x^2 = 8x^3 - 2x^2 - 4$

3) *Simplify the expressions.* $(8x^2 - 3x^3) - (2x^2 + 5x^3) =$

Solution: First use Distributive Property: $-(2x^2 + 5x^3) = -2x^2 - 5x^3 \rightarrow$

$(8x^2 - 3x^3) - (2x^2 + 5x^3) = 8x^2 - 3x^3 - 2x^2 - 5x^3$

Now combine like terms and write in standard form: $8x^2 - 3x^3 - 2x^2 - 5x^3 = -8x^3 + 6x^2$

Multiplying Monomials

- A monomial is a polynomial with just one term: Examples: $2x$ or $7y^2$.
- When you multiply monomials, first multiply the coefficients (a number placed before and multiplying the variable) and then multiply the variables using multiplication property of exponents.

$$x^a \times x^b = x^{a+b}$$

Examples:

1) *Multiply expressions.* $5xy^4z^2 \times 3x^2y^5z^3$

Solution: Find same variables and use multiplication property of exponents: $x^a \times x^b = x^{a+b}$

$x \times x^2 = x^{1+2} = x^3$, $y^4 \times y^5 = y^{4+5} = y^9$ and $z^2 \times z^3 = z^{2+3} = z^5$

Then, multiply coefficients and variables: $5xy^4z^2 \times 3x^2y^5z^3 = 15x^3y^9z^5$

2) *Multiply expressions.* $-2a^5b^4 \times 8a^3b^4 =$

Solution: Use multiplication property of exponents: $x^a \times x^b = x^{a+b}$

$a^5 \times a^3 = a^{5+3} = a^8$ and $b^4 \times b^4 = b^{4+4} = b^8$

Then: $-2a^5b^4 \times 8a^3b^4 = -16a^8b^8$

3) *Multiply.* $7xy^3z^5 \times 4x^2y^4z^3$

Solution: Use multiplication property of exponents: $x^a \times x^b = x^{a+b}$

$x \times x^2 = x^{1+2} = x^3$, $y^3 \times y^4 = y^{3+4} = y^7$ and $z^5 \times z^3 = z^{5+3} = z^8$

Then: $7xy^3z^5 \times 4x^2y^5z^3 = 28x^3y^7z^8$

4) *Simplify.* $(5a^6b^3)(-9a^7b^2) =$

Solution: Use multiplication property of exponents: $x^a \times x^b = x^{a+b}$

$a^6 \times a^7 = a^{6+7} = a^{13}$ and $b^3 \times b^2 = b^{3+2} = b^5$

Then: $(5a^6b^3) \times (-9a^6b^2) = -45a^{13}b^5$

Multiplying and Dividing Monomials

- When you divide or multiply two monomials you need to divide or multiply their coefficients and then divide or multiply their variables.
- In case of exponents with the same base, for Division, subtract their powers, for Multiplication, add their powers.
- Exponent's Multiplication and Division rules:

$$x^a \times x^b = x^{a+b}, \quad \frac{x^a}{x^b} = x^{a-b}$$

Examples:

1) *Multiply expressions.* $(-5x^8)(4x^6) =$

Solution: Use multiplication property of exponents: $x^a \times x^b = x^{a+b} \rightarrow x^8 \times x^6 = x^{14}$

Then: $(-5x^5)(4x^4) = -20x^{14}$

2) *Divide expressions.* $\frac{14x^5y^4}{2xy^3} =$

Solution: Use division property of exponents: $\frac{x^a}{x^b} = x^{a-b} \rightarrow \frac{x^5}{x} = x^{5-1} = x^4$ and $\frac{y^4}{y^3} = y$

Then: $\frac{14x^5y^4}{2xy^3} = 7x^4y$

3) *Divide expressions.* $\frac{56a^8b^3}{8ab^3}$

Solution: Use division property of exponents: $\frac{x^a}{x^b} = x^{a-b} \rightarrow \frac{a^8}{a} = a^{8-1} = a^7$ and $\frac{b^3}{b^3} = 1$

Then: $\frac{56a^8b^3}{8ab^3} = 7a^7$

Multiplying a Polynomial and a Monomial

- When multiplying monomials, use the product rule for exponents.

$$x^a \times x^b = x^{a+b}$$

- When multiplying a monomial by a polynomial, use the distributive property.

$$a \times (b + c) = a \times b + a \times c = ab + ac$$
$$a \times (b - c) = a \times b - a \times c = ab - ac$$

Examples:

1) *Multiply expressions.* $5x(3x - 2)$

Solution: Use Distributive Property: $5x(3x - 2) = 5x \times 3x - 5x \times (-2) = 15x^2 - 10x$

2) *Multiply expressions.* $x(2x^2 + 3y^2)$

Solution: Use Distributive Property: $x(2x^2 + 3y^2) = x \times 2x^2 + x \times 3y^2 = 2x^3 + 3xy^2$

3) *Multiply.* $-4x(-5x^2 + 3x - 6)$

Solution: Use Distributive Property:

$-4x(-5x^2 + 3x - 6) = (-4x)(-5x^2) + (-4x) \times (3x) + (-4x) \times (-6) =$

Now, simplify: $(-4x)(-5x^2) + (-4x) \times (3x) + (-4x) \times (-6) = 20x^3 - 12x^2 + 24x$

Multiplying Binomials

- A binomial is a polynomial that is the sum or the difference of two terms, each of which is a monomial.
- To multiply two binomials, use "FOIL" method. (First-Out-In-Last)

$$(x+a)(x+b) = x \times x + x \times b + a \times x + a \times b = x^2 + bx + ax + ab$$

Examples:

1) *Multiply Binomials.* $(x+2)(x-4) =$

Solution: Use "FOIL". (First–Out–In–Last): $(x+2)(x-4) = x^2 - 4x + 2x - 8$

Then combine like terms: $x^2 - 4x + 2x - 8 = x^2 - 2x - 8$

2) *Multiply.* $(x-5)(x-2) =$

Solution: Use "FOIL". (First–Out–In–Last): $(x-5)(x-2) = x^2 - 2x - 5x + 10$

Then simplify: $x^2 - 2x - 5x + 10 = x^2 - 7x + 10$

3) *Multiply.* $(x-3)(x+6) =$

Solution: Use "FOIL". (First–Out–In–Last): $(x-3)(x+6) = x^2 + 6x - 3x - 18$

Then simplify: $x^2 + 6x - 3x - 18 = x^2 + 3x - 18$

4) *Multiply Binomials.* $(x+8)(x+4) =$

Solution: Use "FOIL". (First–Out–In–Last): $(x+8)(x+4) = x^2 + 4x + 8x + 32$

Then combine like terms: $x^2 + 4x + 8x + 32 = x^2 + 12x + 32$

Factoring Trinomials

To factor trinomials, you can use following methods:

✓ "FOIL": $(x+a)(x+b) = x^2 + (b+a)x + ab$

✓ "Difference of Squares":

$$a^2 - b^2 = (a+b)(a-b)$$

$$a^2 + 2ab + b^2 = (a+b)(a+b)$$

$$a^2 - 2ab + b^2 = (a-b)(a-b)$$

✓ "Reverse FOIL": $x^2 + (b+a)x + ab = (x+a)(x+b)$

Examples:

1) ***Factor this trinomial.*** $x^2 - 2x - 8$

Solution: Break the expression into groups. You need to find two numbers that their product is -8 and their sum is -2. (remember "Reverse FOIL": $x^2 + (b+a)x + ab = (x+a)(x+b)$). Those two numbers are 2 and -4. Then:

$x^2 - 2x - 8 = (x^2 + 2x) + (-4x - 8)$

Now factor out x from $x^2 + 2x$: $x(x+2)$, and factor out -4 from $-4x - 8$: $-4(x+2)$

Then: $(x^2 + 2x) + (-4x - 8) = x(x+2) - 4(x+2)$

Now factor out like term: $(x+2)$. Then: $(x+2)(x-4)$

2) ***Factor this trinomial.*** $x^2 - 2x - 24$

Solution: Break the expression into groups: $(x^2 + 4x) + (-6x - 24)$

Now factor out x from $x^2 + 4x$: $x(x+4)$, and factor out -6 from $-6x - 24$: $-6(x+4)$

Then: $(x+4) - 6(x+4)$, now factor out like term:

$(x = 4) \rightarrow x(x+4) - 6(x+4) = (x+4)(x-6)$

Chapter 10: Practices

Simplify each polynomial.

1) $4(3x+5) =$

2) $8(6x-1) =$

3) $2x(7x+4)+2x =$

4) $5x(6x+2)-4x =$

5) $7x(3x-6)+x^2-2 =$

6) $3x^2-5-9x(4x+1) =$

Add or subtract polynomials.

7) $(6x^2+8)+(2x^2-4) =$

8) $(x^2-5x)-(3x^2+2x) =$

9) $(9x^3-4x^2)+(x^3-3x^2) =$

10) $(7x^3-3x)-(6x^3-5x) =$

11) $(12x^3+x^2)+(2x^2-10) =$

12) $(4x^3-15)-(3x^3-7x^2) =$

Simplify each expression. (Multiplying Monomials)

13) $5x^3 \times 6x^4 =$

14) $-4a^2b \times 3ab^2 =$

15) $(-7x^2yz) \times (-3xy^3z^2) =$

16) $9u^4t^2 \times (-4ut) =$

17) $12x^3z \times 3xy^2 =$

18) $-8a^2bc \times a^3b^2 =$

Simplify each expression. (Multiplying and Dividing Monomials)

19) $(4x^3y^4)(12x^5y^2) =$

20) $(5x^4y^2)(8x^6y^5) =$

21) $(16x^7y^9)(3x^6y^4) =$

22) $\frac{36x^5y^3}{9x^2y} =$

23) $\frac{98x^{12}y^{10}}{7x^9y^7} =$

24) $\frac{22\quad ^{9}y^{13}}{15\quad ^{6}y^{9}} =$

Find each product. (Multiplying a Polynomial and a Monomial)

25) $4x(6x - y) =$

26) $7x(3x + 5y) =$

27) $5x(x - 8y) =$

28) $x(3x^2 + 2x - 6) =$

29) $5x(-x^2 + 7x + 4) =$

30) $6x(6x^2 - 3x - 12) =$

Find each product. (Multiplying Binomials)

31) $(x - 3)(x + 3) =$

32) $(x - 5)(x - 4) =$

33) $(x + 6)(x + 3) =$

34) $(x - 7)(x + 8) =$

35) $(x + 2)(x - 9) =$

36) $(x - 15)(x + 3) =$

Factor each trinomial.

37) $x^2 + 4x - 12 =$

38) $x^2 + x - 20 =$

39) $x^2 + 3x - 108 =$

40) $x^2 + 12x + 32 =$

41) $x^2 - 14x + 48 =$

42) $x^2 + 2x - 35 =$

Answers – Chapter 10

1) $12x + 20$
2) $48x - 8$
3) $14x^2 + 10x$

4) $30x^2 + 6x$
5) $22x^2 - 42x - 2$
6) $-33x^2 - 9x - 5$

7) $8x^2 + 4$
8) $-2x^2 - 7x$
9) $10x^3 - 7x^2$

10) $x^3 + 2x$
11) $12x^3 + 3x^2 - 10$
12) $x^3 + 7x^2 - 15$

13) $30x^7$
14) $-12a^3b^3$
15) $21x^3y^4z^3$

16) $-36u^5t^3$
17) $36x^4y^2\text{z}$
18) $-8a^5b^3\text{c}$

19) $48x^8y^6$
20) $40x^{10}y^7$
21) $48x^{13}y^{13}$
22) $4x^3y^2$

23) $14x^3y^3$

24) $15x^3y^4$

25) $24x^2 - 4xy$
26) $21x^2 + 35xy$
27) $5x^2 - 40xy$

28) $3x^3 + 2x^2 - 6x$
29) $-5x^3 + 35x^2 + 20x$
30) $36x^3 - 18x^2 - 72x$

31) $x^2 - 9$
32) $x^2 - 9x + 20$
33) $x^2 + 9x + 18$

34) $x^2 + x - 56$
35) $x^2 - 7x - 18$
36) $x^2 - 12x - 45$

37) $(x - 2)(x + 6)$
38) $(x + 5)(x - 4)$
39) $(x + 12)(x - 9)$

40) $(x + 8)(x + 4)$
41) $(x - 6)(x - 8)$
42) $(x + 7)(x - 5)$

Chapter 11:

Geometry and Solid Figures

Math Topics that you'll learn in this Chapter:

- ✓ The Pythagorean Theorem
- ✓ Triangles
- ✓ Polygons
- ✓ Circles
- ✓ Trapezoids
- ✓ Cubes
- ✓ Rectangle Prisms
- ✓ Cylinder

The Pythagorean Theorem

- You can use the Pythagorean Theorem to find a missing side in a right triangle.
- In any right triangle: $a^2 + b^2 = c^2$

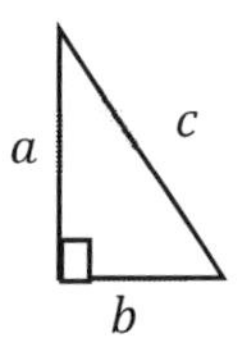

Examples:

1) Right triangle ABC (not shown) has two legs of lengths 6 cm (AB) and 8 cm (AC). What is the length of the hypotenuse of the triangle (side BC)?

Solution: Use Pythagorean Theorem: $a^2 + b^2 = c^2$, $a = 6$, and $b = 8$

Then: $a^2 + b^2 = c^2 \rightarrow 6^2 + 8^2 = c^2 \rightarrow 36 + 64 = c^2 \rightarrow 100 = c^2 \rightarrow c = \sqrt{100} = 10$

The length of the hypotenuse is 10 cm.

2) Find the hypotenuse of the following triangle.

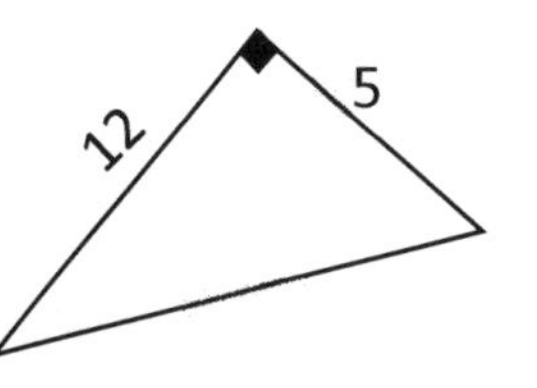

Solution: Use Pythagorean Theorem: $a^2 + b^2 = c^2$

Then: $a^2 + b^2 = c^2 \rightarrow 12^2 + 5^2 = c^2 \rightarrow 144 + 25 = c^2$

$c^2 = 169 \rightarrow c = \sqrt{169} = 13$

3) Find the length of the missing side in the following triangle.

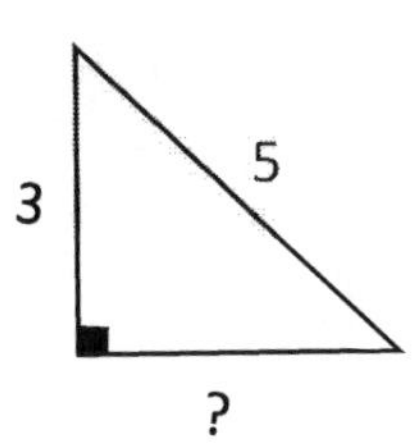

Solution: Use Pythagorean Theorem: $a^2 + b^2 = c^2$

Then: $a^2 + b^2 = c^2 \rightarrow 3^2 + b^2 = 5^2 \rightarrow 9 + b^2 = 25 \rightarrow$

$b^2 = 25 - 9 \rightarrow b^2 = 16 \rightarrow b = \sqrt{16} = 4$

Triangles

- In any triangle the sum of all angles is 180 degrees.
- Area of a triangle $= \frac{1}{2}\,(base \times height)$

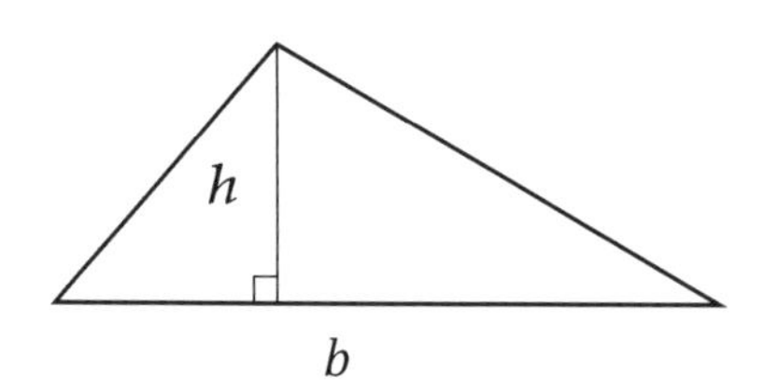

Examples:

What is the area of following triangles?

1)

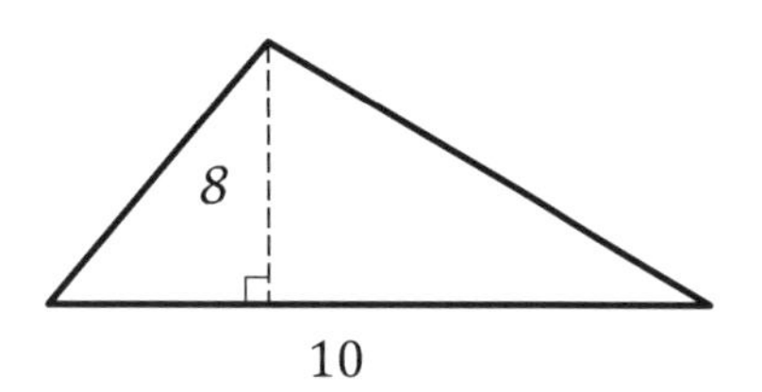

Solution:

Use the area formula: Area $= \frac{1}{2}\,(base \times height)$

$base = 10$ and $height = 8$

Area $= \frac{1}{2}(10 \times 8) = \frac{1}{2}(80) = 40$

2)

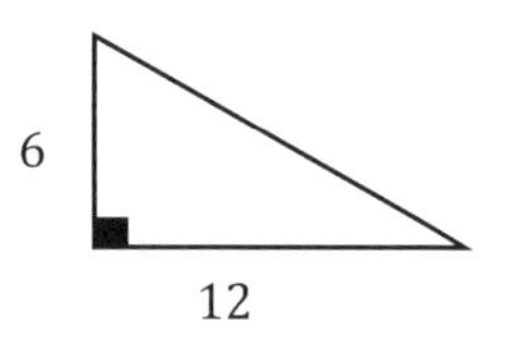

Solution:

Use the area formula: Area $= \frac{1}{2}\,(base \times height)$

$base = 12$ and $height = 6$

Area $= \frac{1}{2}(12 \times 6) = \frac{72}{2} = 36$

3) What is the missing angle in the following triangle?

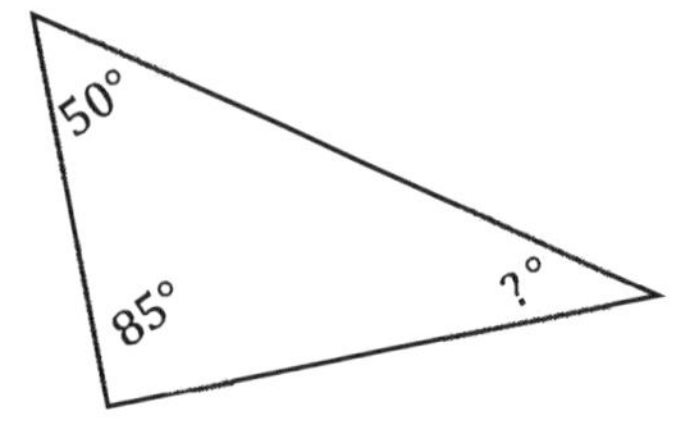

Solution:

In any triangle the sum of all angles is 180 degrees.

Let x be the missing angle. Then: $50 + 85 + x = 180$

$\rightarrow 135 + x = 180 \rightarrow x = 180 - 135 = 45$

The missing angle is 45 degrees.

Polygons

✓ Perimeter of a square

$= 4 \times side = 4s$

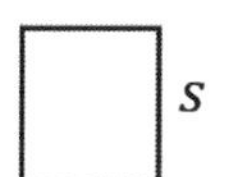

✓ Perimeter of a rectangle

$= 2(width + length)$

width

length

✓ Perimeter of trapezoid

$= a + b + c + d$

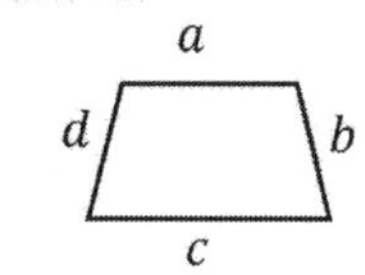

✓ Perimeter of a regular hexagon $= 6a$

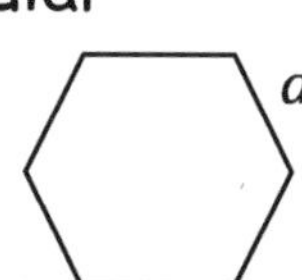

✓ Perimeter of a parallelogram $= 2(l + w)$

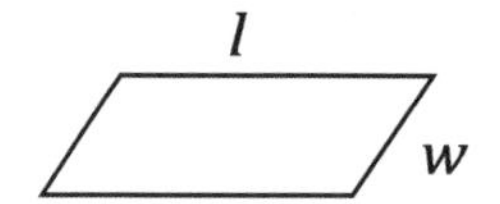

Examples:

1) Find the perimeter of following regular hexagon.

Solution: Since the hexagon is regular, all sides are equal.

Then: Perimeter of Hexagon $= 6 \times (one\ side)$

Perimeter of Hexagon $= 6 \times (one\ side) = 6 \times 4 = 24\ m$

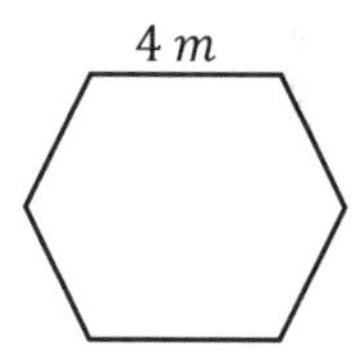

2) Find the perimeter of following trapezoid.

Solution: Perimeter of a trapezoid $= a + b + c + d$

Perimeter of the trapezoid $= 5 + 6 + 6 + 8 = 25\ ft$

5 ft

6 ft

6 ft

8 ft

Circles

- In a circle, variable r is usually used for the radius and d for diameter.
- $Area\ of\ a\ circle = \pi r^2$ (π is about 3.14)
- $Circumference\ of\ a\ circle = 2\pi r$

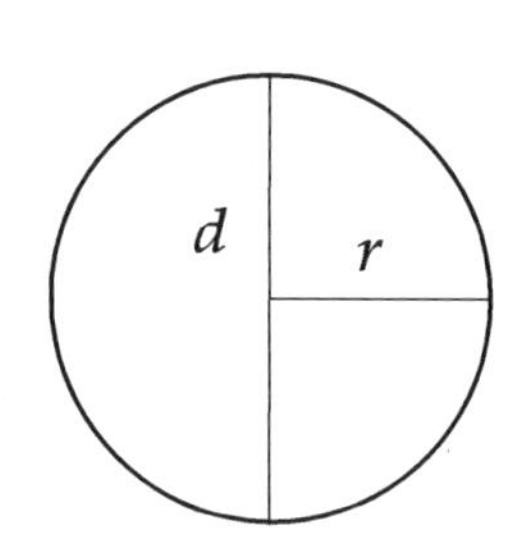

Examples:

1) Find the area of the following circle.

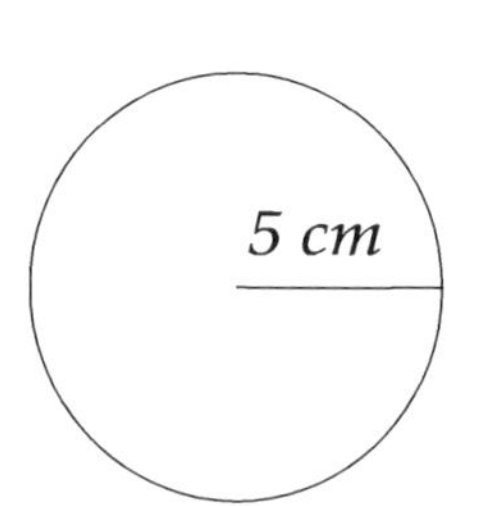

Solution:

Use area formula: $Area = \pi r^2$

$r = 8\ in \rightarrow Area = \pi(8)^2 = 64\pi, \pi = 3.14$

Then: $Area = 64 \times 3.14 = 200.96\ in^2$

2) Find the Circumference of the following circle.

Solution:

Use Circumference formula: $Circumference = 2\pi r$

$r = 5\ cm \rightarrow Circumference = 2\pi(5) = 10\pi$

$\pi = 3.14$ ***Then:*** $Circumference = 10 \times 3.14 = 31.4\ cm$

3) Find the area of the circle.

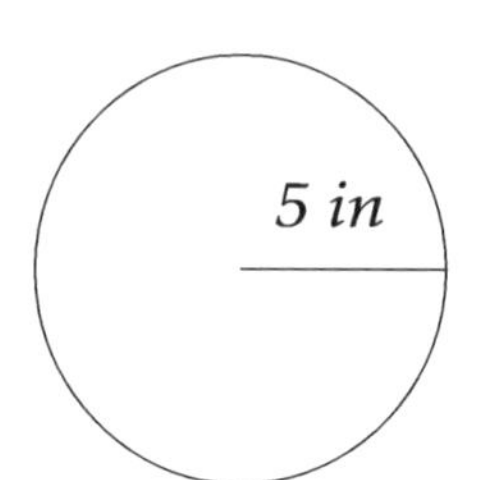

Solution:

Use area formula: $Area = \pi r^2$,

$r = 5\ in$ then: $Area = \pi(5)^2 = 25\pi, \pi = 3.14$

Then: $Area = 25 \times 3.14 = 78.5$

Trapezoids

- A quadrilateral with at least one pair of parallel sides is a trapezoid.
- Area of a trapezoid $= \frac{1}{2}h(b_1 + b_2)$

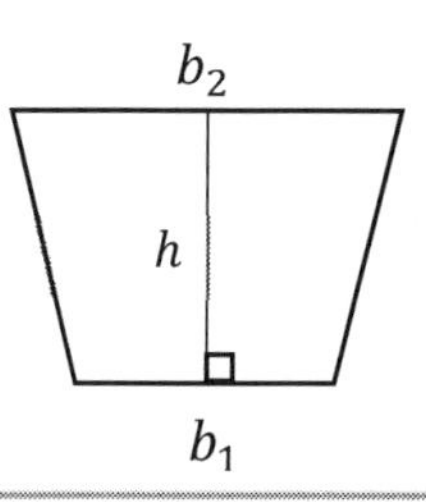

Examples:

1) Calculate the area of the following trapezoid.

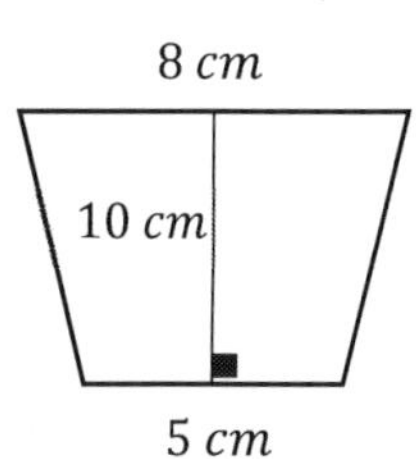

Solution:

Use area formula: $A = \frac{1}{2}h(b_1 + b_2)$

$b_1 = 5\ cm$, $b_2 = 8\ cm$ and $h = 10\ cm$

Then: $A = \frac{1}{2}(10)(8 + 5) = 5(13) = 65\ cm^2$

2) Calculate the area of the following trapezoid.

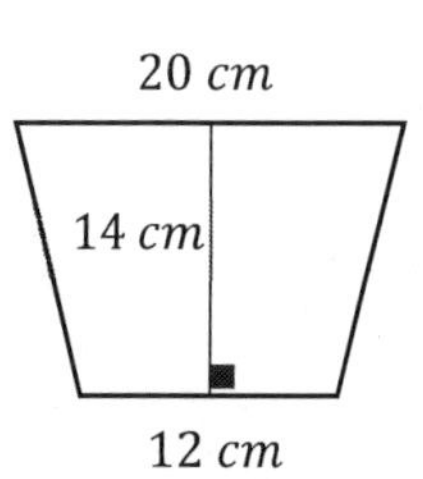

Solution:

Use area formula: $A = \frac{1}{2}h(b_1 + b_2)$

$b_1 = 12\ cm$, $b_2 = 20\ cm$ and $h = 14\ cm$

Then: $A = \frac{1}{2}(14)(12 + 20) = 7(32) = 224\ cm^2$

Cubes

- A cube is a three-dimensional solid object bounded by six square sides.
- Volume is the measure of the amount of space inside of a solid figure, like a cube, ball, cylinder or pyramid.
- Volume of a cube $= (\boldsymbol{one\ side})^3$
- surface area of a cube $= \boldsymbol{6} \times (\boldsymbol{one\ side})^2$

Examples:

1) Find the volume and surface area of the following cube.

Solution: Use volume formula: $volume = (one\ side)^3$

Then: $volume = (one\ side)^3 = (2)^3 = 8\ cm^3$

Use surface area formula: $surface\ area\ of\ cube: 6(one\ side)^2 =$

$$6(2)^2 = 6(4) = 24\ cm^2$$

2 cm

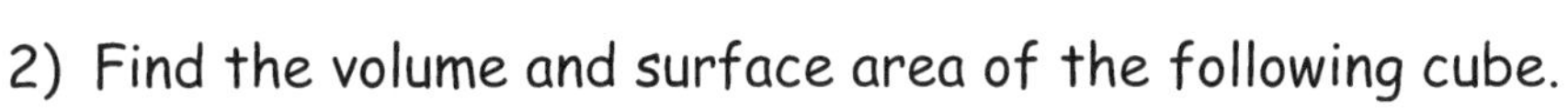

2) Find the volume and surface area of the following cube.

Solution: Use volume formula: $volume = (one\ side)^3$

Then: $volume = (one\ side)^3 = (5)^3 = 125\ cm^3$

Use surface area formula:

$surface\ area\ of\ cube: 6(one\ side)^2 = 6(5)^2 = 6(25) = 150\ cm^2$

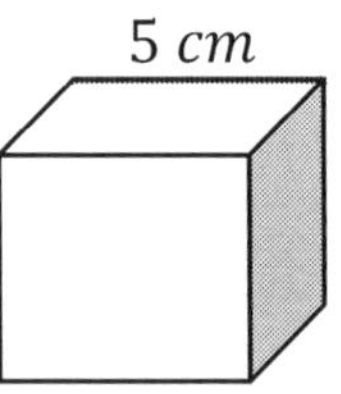

3) Find the volume and surface area of the following cube.

Solution: Use volume formula: $volume = (one\ side)^3$

Then: $volume = (one\ side)^3 = (7)^3 = 343\ m^3$

Use surface area formula:

$surface\ area\ of\ cube: 6(one\ side)^2 = 6(7)^2 = 6(49) = 294\ m^2$

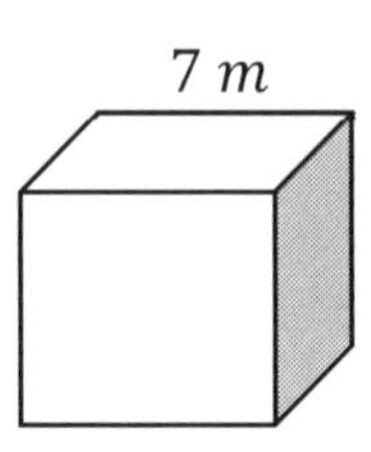

Rectangular Prisms

- A rectangular prism is a solid 3-dimensional object which has six rectangular faces.
- Volume of a Rectangular prism $= \boldsymbol{Length} \times \boldsymbol{Width} \times \boldsymbol{Height}$

$Volume = l \times w \times h$

$Surface\ area = 2 \times (wh + lw + lh)$

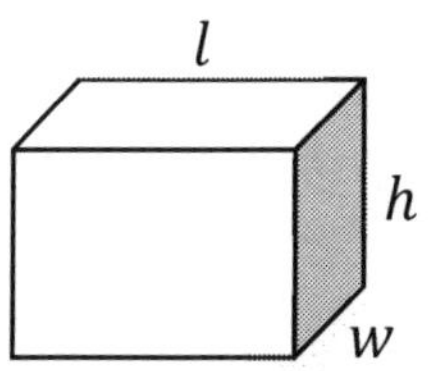

Examples:

1) Find the volume and surface area of the following rectangular prism.

8 m
10 m
6 m

Solution:

Use volume formula: $Volume = l \times w \times h$

Then: $Volume = 8 \times 6 \times 10 = 480\ m^3$

Use surface area formula: $Surface\ area = 2 \times (wh + lw + lh)$

Then: $Surface\ area = 2 \times \big((6 \times 10) + (8 \times 6) + (8 \times 10)\big)$

$= 2 \times (60 + 48 + 80) = 2 \times (188) = 376\ m^2$

2) Find the volume and surface area of rectangular prism.

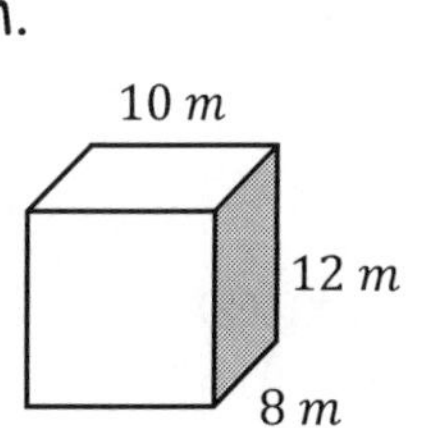

Solution:

Use volume formula: $Volume = l \times w \times h$

Then: $Volume = 10 \times 8 \times 12 = 960\ m^3$

Use surface area formula: $Surface\ area = 2 \times (wh + lw + lh)$

Then: $Surface\ area = 2 \times \big((8 \times 12) + (10 \times 8) + (10 \times 12)\big)$

$= 2 \times (96 + 80 + 120) = 2 \times (296) = 592\ m^2$

Cylinder

- A cylinder is a solid geometric figure with straight parallel sides and a circular or oval cross section.
- $Volume\ of\ a\ Cylinder = \pi(radius)^2 \times height,\ \pi \approx 3.14$
- $Surface\ area\ of\ a\ cylinder = 2\pi r^2 + 2\pi rh$

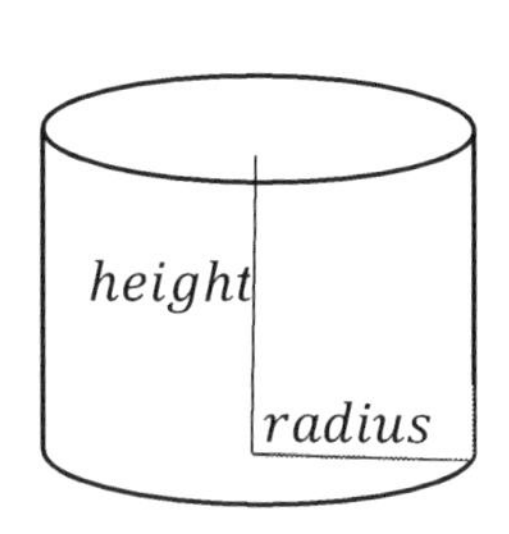

Examples:

1) *Find the volume and Surface area of the follow Cylinder.*

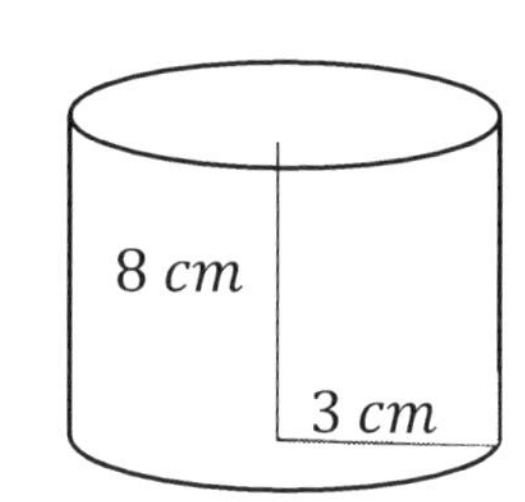

Solution:

Use volume formula: $Volume = \pi(radius)^2 \times height$
Then: $Volume = \pi(3)^2 \times 8 = 9\pi \times 8 = 72\pi$
$\pi = 3.14$ ***then:*** $Volume = 72\pi = 72 \times 3.14 = 226.08\ cm^3$
Use surface area formula: $Surface\ area = 2\pi r^2 + 2\pi rh$
Then: $\mathbf{2}\pi(3)^2 + 2\pi(3)(8) = 2\pi(9) + 2\pi(24) = 18\pi + 48\pi = 66\pi$
$\pi = 3.14$ Then: $Surface\ area = 66 \times 3.14 = 207.24\ cm^2$

2) *Find the volume and Surface area of the follow Cylinder.*

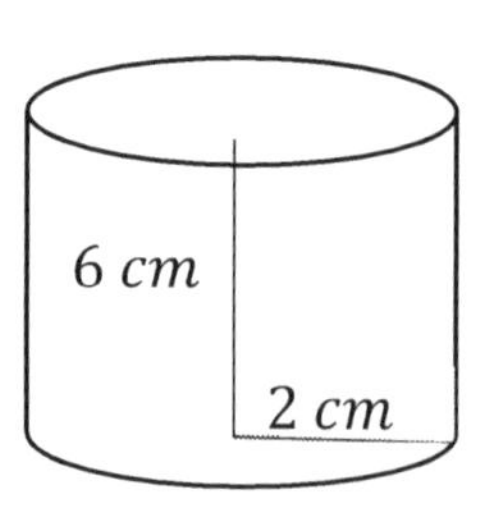

Solution:

Use volume formula: $Volume = \pi(radius)^2 \times height$
Then: $Volume = \pi(2)^2 \times 6 = \pi 4 \times 6 = 24\pi$
$\pi = 3.14$ ***then:*** $Volume = 24\pi = 75.36\ cm^3$
Use surface area formula: $Surface\ area = 2\pi r^2 + 2\pi rh$
Then: $= 2\pi(2)^2 + 2\pi(2)(6) = 2\pi(4) + 2\pi(12) = 8\pi + 24\pi = 32\pi$
$\pi = 3.14$ ***then:*** $Surface\ area = 32 \times 3.14 = 100.48\ cm^2$

Chapter 11: Practices

Find the missing side?

1)

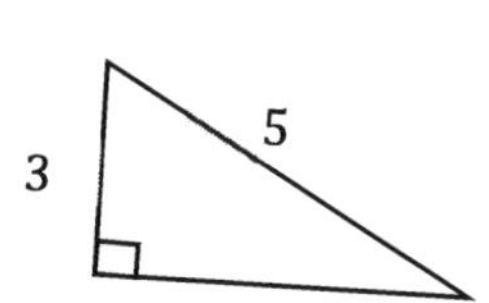

2)

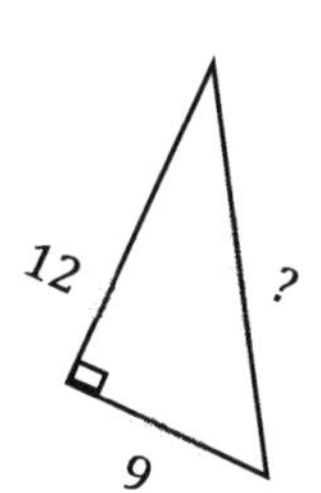

3)

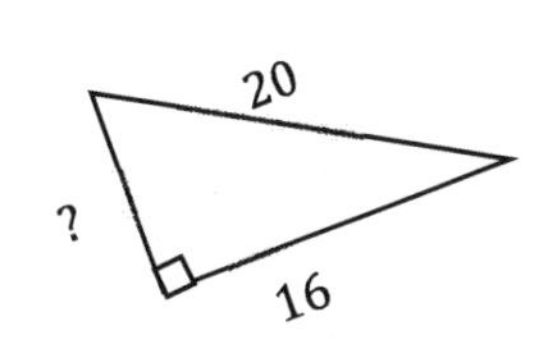

4)

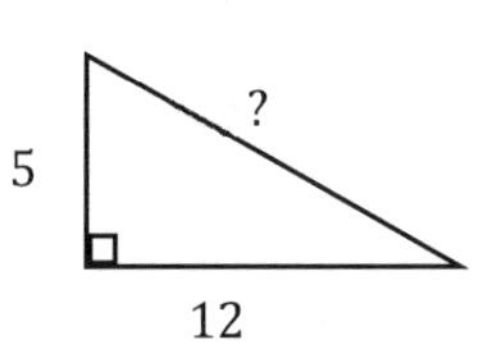

Find the measure of the unknown angle in each triangle.

5)

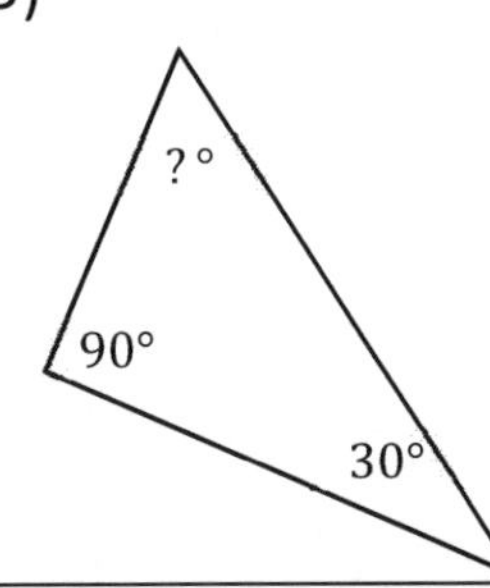

6)

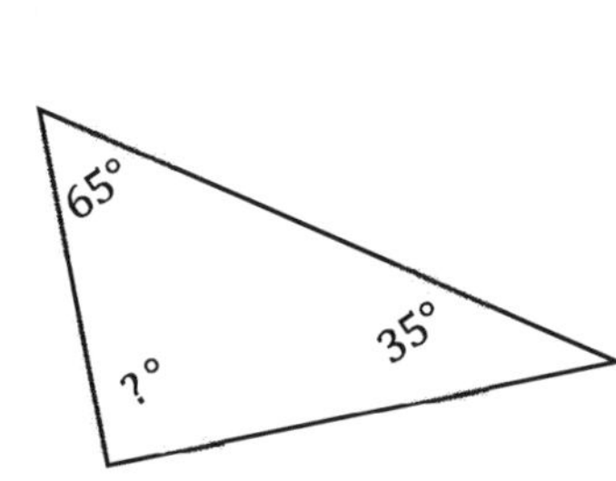

7)

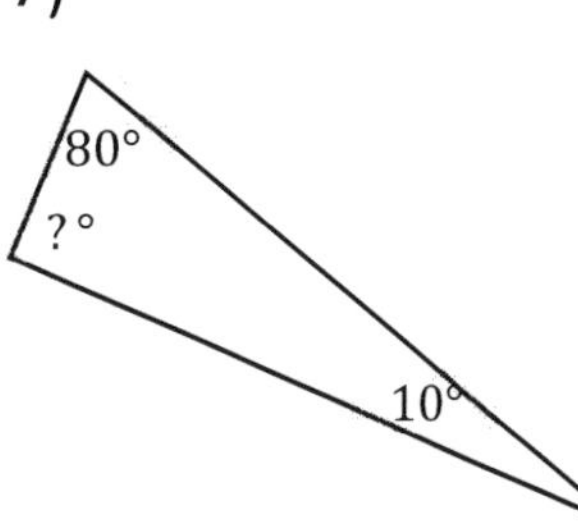

8)

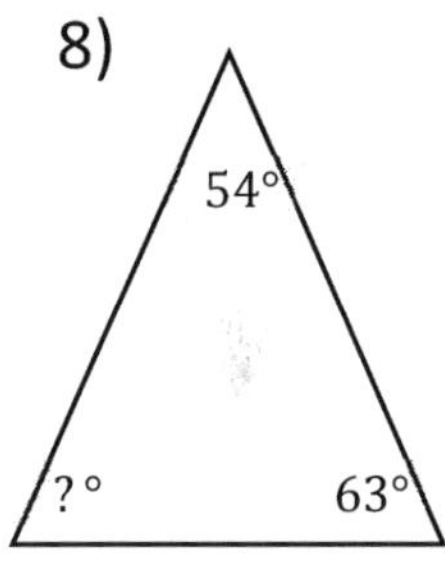

Find area of each triangle.

9)

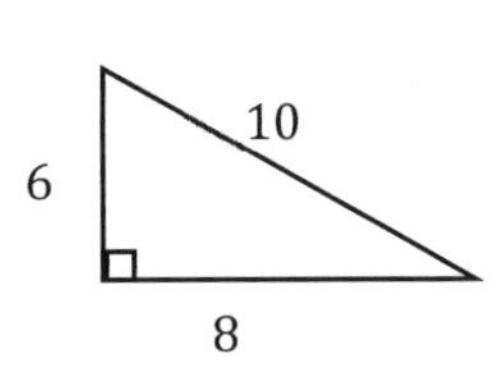

10)

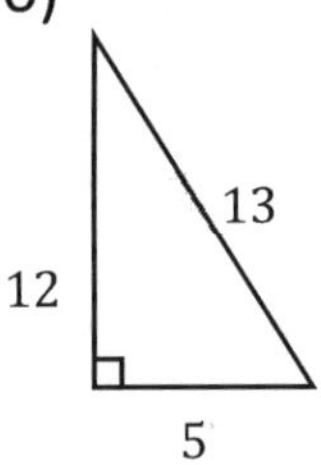

11)

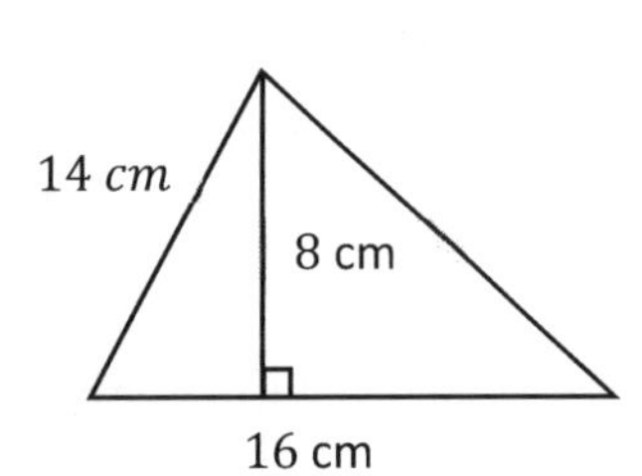

12)

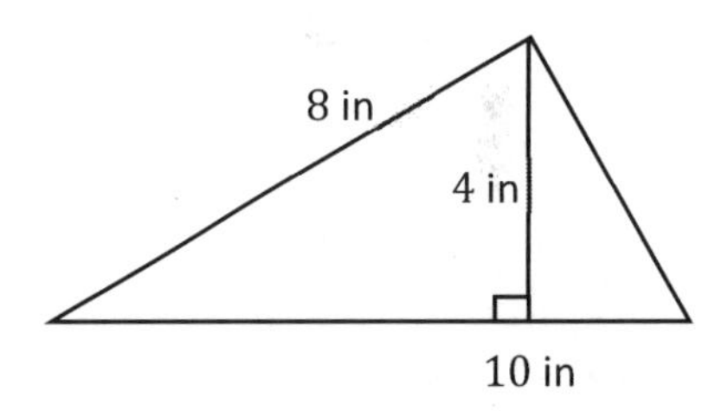

Find the perimeter of each shape.

13)

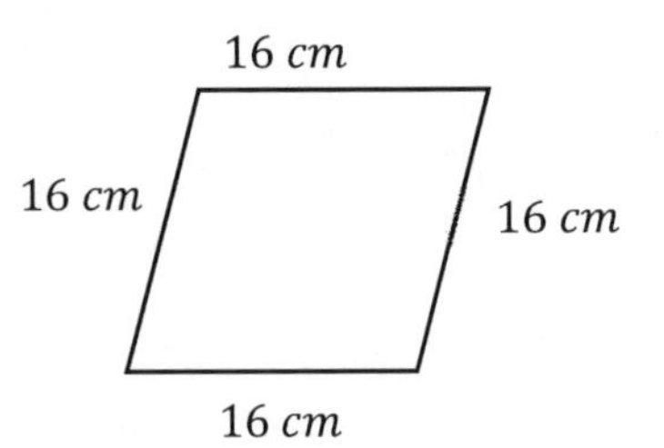

14)

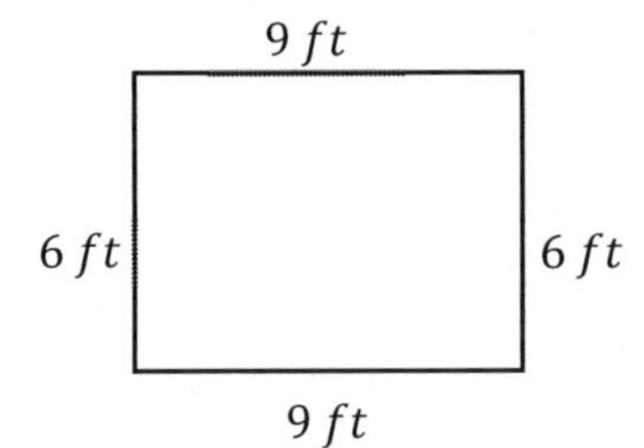

15)

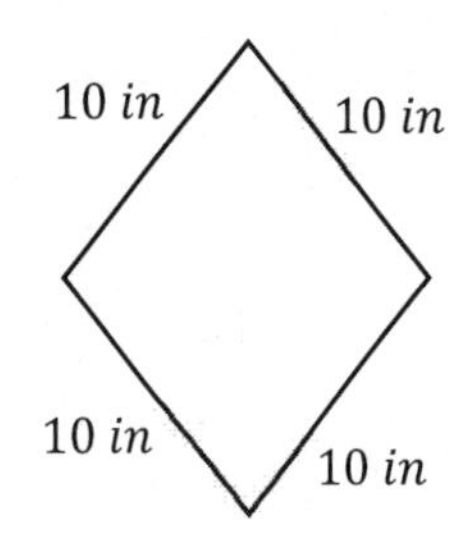

16)

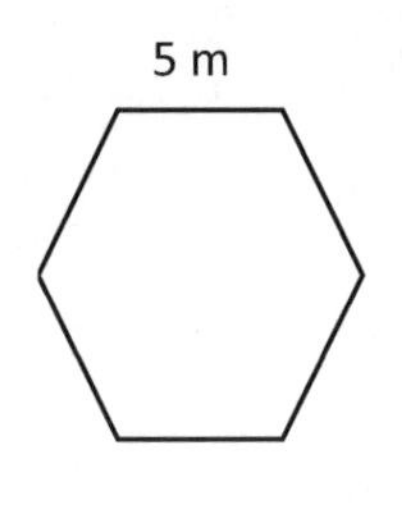

Complete the table below. $(\pi = 3.14)$

17)

	Radius	Diameter	Circumference	Area
Circle 1	3 *inches*	6 *inches*	18.84 *inches*	28.26 *square inches*
Circle 2			43.96 *meters*	
Circle 3		8 *ft*		
Circle 4				78.5 *square miles*

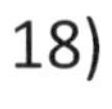

Find the area of each trapezoid.

18)

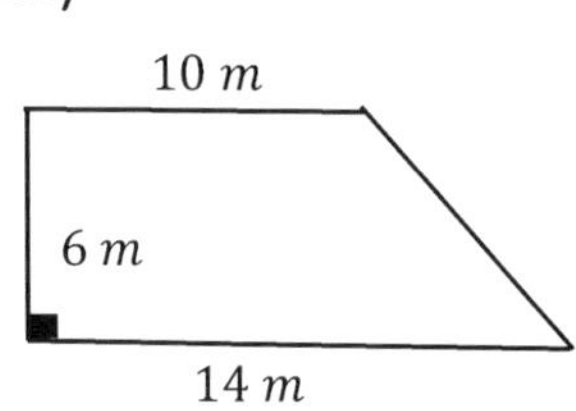

19)

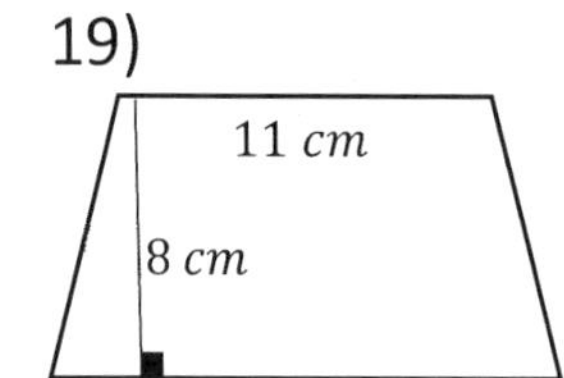

20)

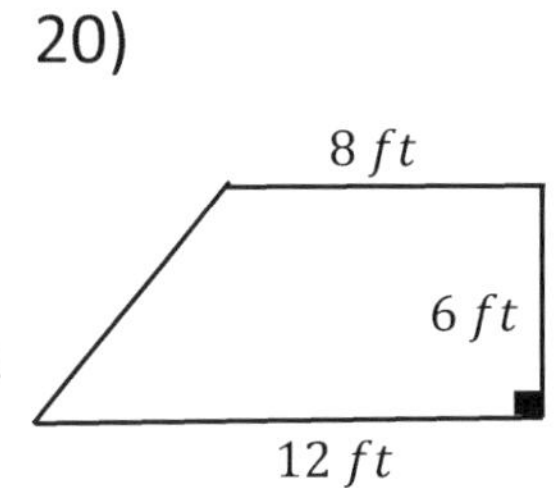

21)

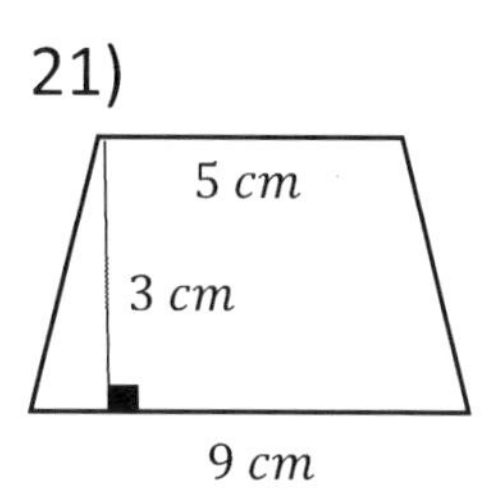

Find the volume of each cube.

22)

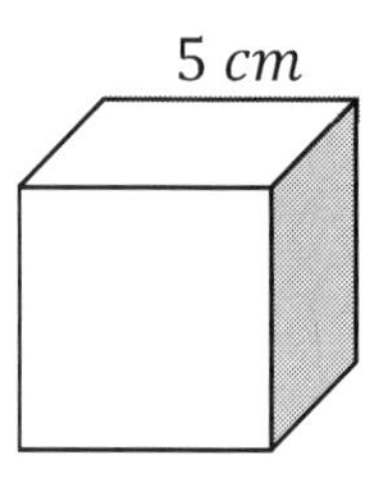

23)

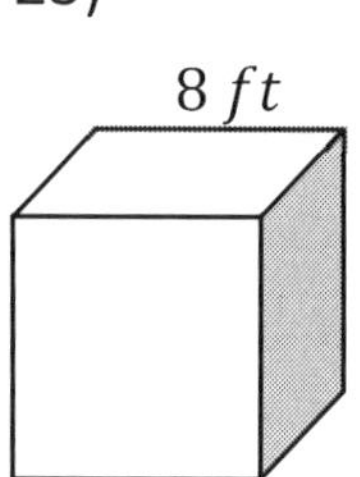

24)

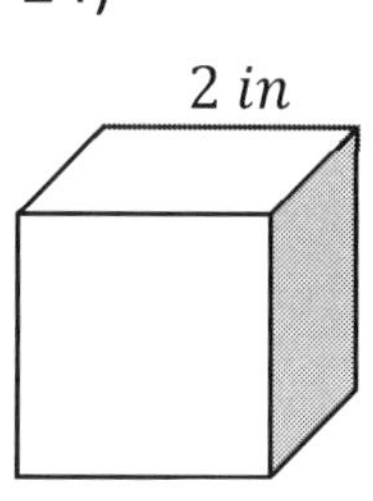

25)

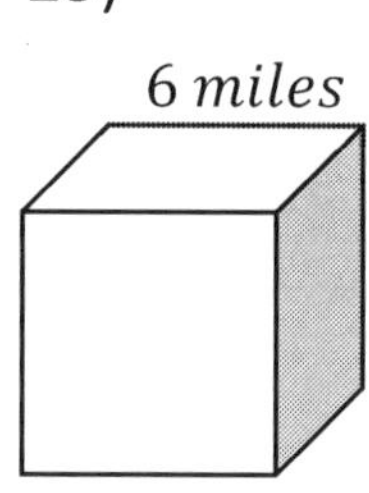

Find the volume of each Rectangular Prism.

26)

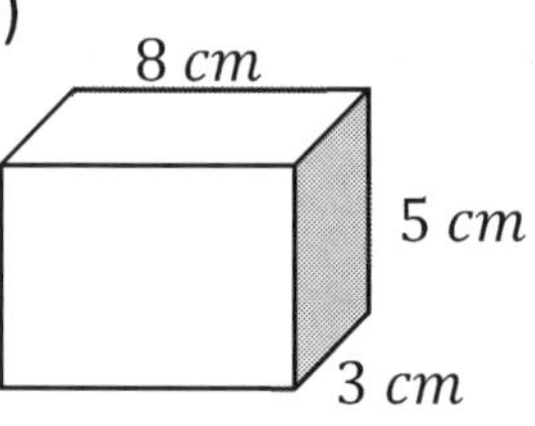

27)

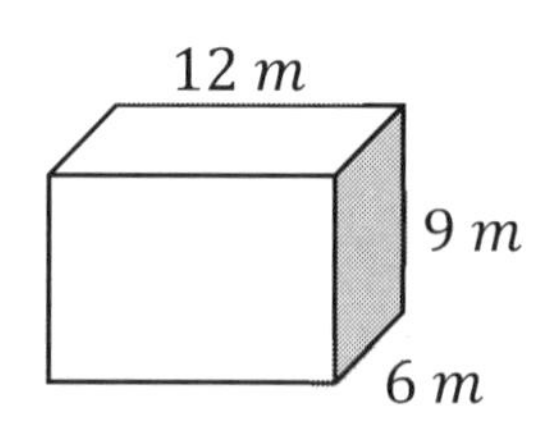

28)

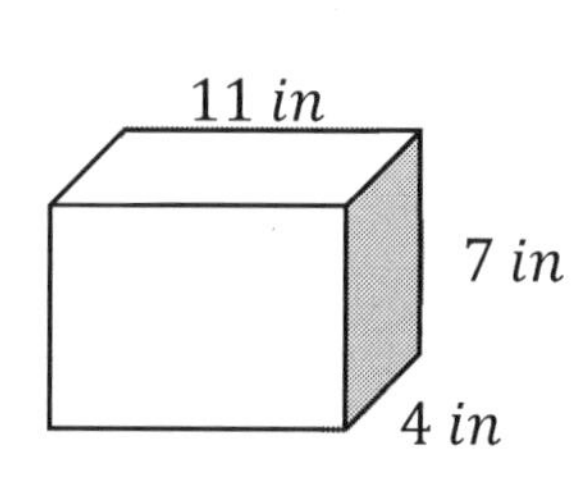

Find the volume of each Cylinder. Round your answer to the nearest tenth. ($\pi = 3.14$)

25)

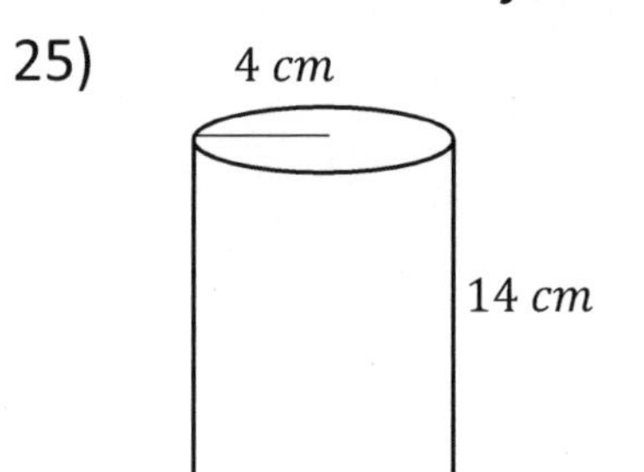

26)

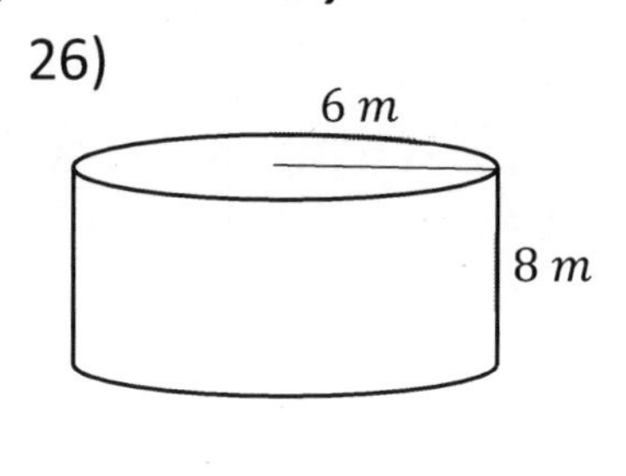

27)

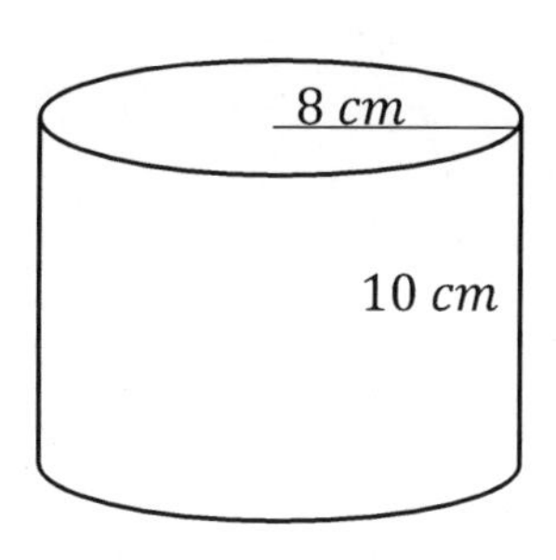

Answers – Chapter 11

1) 4
2) 15
3) 12
4) 13

5) $60°$
6) $80°$
7) $90°$
8) $63°$
9) 24 *square unites*
10) 30 *square unites*
11) $64\ cm^2$
12) $20\ in^2$

13) $64\ cm$
14) $30\ ft$
15) $40\ in$
16) $30\ m$

17)

	Radius	Diameter	Circumference	Area
Circle 1	3 *inches*	6 *inches*	18.84 *inches*	28.26 *square inches*
Circle 2	7 *meters*	14 *meters*	43.96 *meters*	153.86 *square meters*
Circle 3	4 *ft*	8 *ft*	25.12 *ft*	50.24 *square ft*
Circle 4	5 *miles*	10 *miles*	31.4 *miles*	78.5 *square miles*

18) $72\ m^2$
19) $108\ cm^2$
20) $60\ ft^2$
21) $21\ cm^2$

22) $125\ cm^3$
23) $512\ ft^3$
24) $8\ in^3$
25) $216\ miles^3$

26) $120\ cm^3$
27) $648\ m^3$
28) $308\ in^3$

29) $703.36\ cm^3$
30) $904.32\ m^3$
31) $2{,}009.6\ cm^3$

Chapter 12:

Statistics

Math Topics that you'll learn in this Chapter:

- ✓ Mean, Median, Mode, and Range of the Given Data
- ✓ Pie Graph
- ✓ Probability Problems
- ✓ Permutations and Combinations

Mean, Median, Mode, and Range of the Given Data

- Mean: $\frac{sum\ of\ the\ data}{total\ number\ of\ data\ entires}$
- Mode: the value in the list that appears most often
- Median: is the middle number of a group of numbers that have been arranged in order by size.
- Range: the difference of largest value and smallest value in the list

Examples:

1) What is the mode of these numbers? $4, 5, 7, 5, 7, 4, 0, 4$

Solution: Mode: the value in the list that appears most often.
Therefore, the mode is number 4. There are three number 4 in the data.

2) What is the median of these numbers? $5, 10, 14, 9, 16, 19, 6$

Solution: Write the numbers in order: $5, 6, 9, 10, 14, 16, 19$

Median is the number in the middle. Therefore, the median is 10.

3) What is the mean of these numbers? $8, 2, 8, 5, 3, 2, 4, 8$

Solution: Mean: $\frac{sum\ of\ the\ data}{total\ number\ of\ data\ entires} = \frac{8+2+8+5+3+2+4+8}{8} = 5$

4) What is the range in this list? $4, 9, 13, 8, 15, 18, 5$

Solution: Range is the difference of largest value and smallest value in the list. The largest value is 18 and the smallest value is 4. Then: $18 - 4 = 14$

Pie Graph

- ✓ A Pie Chart is a circle chart divided into sectors, each sector represents the relative size of each value.
- ✓ Pie charts represent a snapshot of how a group is broken down into smaller pieces.

Example:

A library has 820 books that include Mathematics, Physics, Chemistry, English and History. Use following graph to answer the questions.

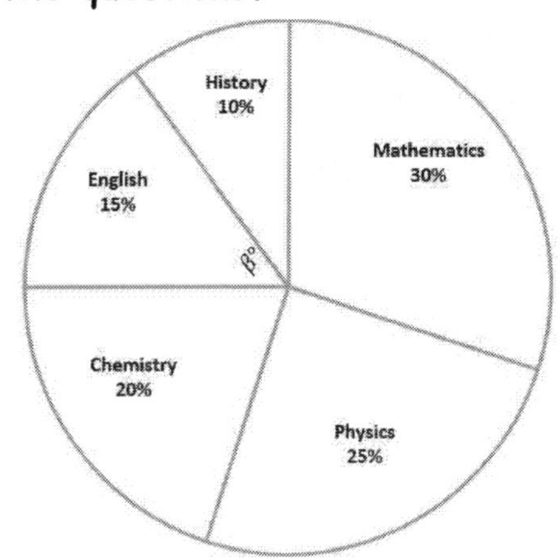

1) What is the number of Mathematics books?

Solution: Number of total books $= 820$

Percent of Mathematics books $= 30\% = 0.30$

Then, number of Mathematics books:

$$0.30 \times 820 = 246$$

2) What is the number of History books?

Solution: Number of total books $= 820$

Percent of History books $= 10\% = 0.10$

Then: $0.10 \times 820 = 82$

3) What is the number of Chemistry books?

Solution: Number of total books $= 820$

Percent of Chemistry books $= 20\% = 0.20$

Then: $0.20 \times 820 = 164$

Probability Problems

- Probability is the likelihood of something happening in the future. It is expressed as a number between zero (can never happen) to 1 (will always happen).
- Probability can be expressed as a fraction, a decimal, or a percent.
- Probability formula: $Probability = \frac{number\ of\ desired\ outcomes}{number\ of\ total\ outcomes}$

1) Anita's trick-or-treat bag contains 12 pieces of chocolate, 18 suckers, 18 pieces of gum, 24 pieces of licorice. If she randomly pulls a piece of candy from her bag, what is the probability of her pulling out a piece of sucker?

Solution: $\text{Probability} = \frac{number\ of\ desired\ outcomes}{\text{number of total outcomes}}$

$$\text{Probability of ulling out a piece of sucker} = \frac{18}{12+18+18+24} = \frac{18}{72} = \frac{1}{4}$$

2) A bag contains 20 balls: four green, five black, eight blue, a brown, a red and one white. If 19 balls are removed from the bag at random, what is the probability that a brown ball has been removed?

Solution: If 19 balls are removed from the bag at random, there will be one ball in the bag. The probability of choosing a brown ball is 1 out of 20. Therefore, the probability of not choosing a brown ball is 19 out of 20 and the probability of having not a brown ball after removing 19 balls is the same.

Permutations and Combinations

- Factorials are products, indicated by an exclamation mark. For example, $4! = 4 \times 3 \times 2 \times 1$(Remember that 0! is defined to be equal to 1.)
- Permutations: The number of ways to choose a sample of k elements from a set of n distinct objects where order does matter, and replacements are not allowed. For a permutation problem, use this formula:

$$_{n}P_{k} = \frac{n!}{(n-k)!}$$

- Combination: The number of ways to choose a sample of r elements from a set of n distinct objects where order does not matter, and replacements are not allowed. For a combination problem, use this formula:

$$_{n}C_{r} = \frac{n!}{r!\,(n-r)!}$$

Examples:

1) *How many ways can the first and second place be awarded to* 8 *people?*

Solution: Since the order matters, (the first and second place are different!) we need to use permutation formula where n is 10 and k is 2. Then: $\frac{n!}{(n-k)!} = \frac{8!}{(8-2)!} = \frac{8!}{6!} = \frac{8\times7\times6!}{6!}$, remove 6! from both sides of the fraction. Then: $\frac{8\times7\times6!}{6!} = 8 \times 7 = 56$

2) *How many ways can we pick a team of* 2 *people from a group of* 6*?*

Solution: Since the order doesn't matter, we need to use combination formula where n is 8 and r is 3. Then: $\frac{n!}{r!\,(n-r)!} = \frac{6!}{2!\,(6-2)!} = \frac{6!}{2!\,(4)!} = \frac{6\times5\times4!}{2!\,(4)!} = \frac{6\times5}{2\times1} = \frac{30}{2} = 15$

Chapter 12: Practices

✍ ***Find the values of the Given Data.***

1) $6, 12, 1, 1, 5$

Mode: _____ Range: _____

Mean: _____ Median: _____

2) $5, 8, 3, 7, 4, 3$

Mode: _____ Range: _____

Mean: _____ Median: _____

✍ The circle graph below shows all Jason's expenses for last month. Jason spent $864 on his bills last month.

3) How much did Jason spend on his car last month? _________

4) How much did Jason spend for foods last month? _________

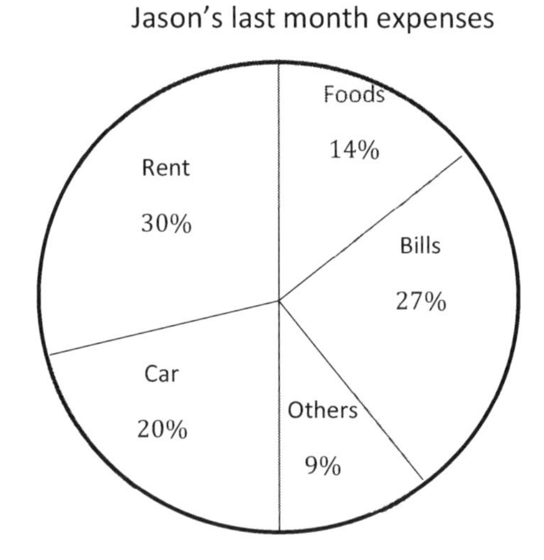

✍ ***Solve.***

5) Bag A contains 8 red marbles and 6 green marbles. Bag B contains 5 black marbles and 10 orange marbles. What is the probability of selecting a green marble at random from bag A? What is the probability of selecting a black marble at random from Bag B? __________ ______________

✍ ***Solve.***

6) Susan is baking cookies. She uses sugar, flour, butter, and eggs. How many different orders of ingredients can she try? ________________

7) Jason is planning for his vacation. He wants to go to museum, watch a movie, go to the beach, and play volleyball. How many different ways of ordering are there for him? ________________

8) In how many ways a team of 10 basketball players can to choose a captain and co-captain? ________________

9) How many ways can you give 3 balls to your 8 friends? ________________

10) A professor is going to arrange her 8 students in a straight line. In how many ways can she do this? ________________

Answers – Chapter 12

1) Mode: 1 , Range: 11, Mean: 5, Median: 5

2) Mode: 3 , Range: 5 , Mean: 5, Median: 4.5

3) $640

4) $448

5) $\frac{3}{7}, \frac{1}{3}$

6) 24

7) 24

8) 90

9) 56

10) 40,320

Chapter 13:

Functions Operations

Math Topics that you'll learn in this Chapter:

- ✓ Function Notation
- ✓ Adding and Subtracting Functions
- ✓ Multiplying and Dividing Functions
- ✓ Composition of Functions

Function Notation and Evaluation

- Functions are mathematical operations that assign unique outputs to given inputs.
- Function notation is the way a function is written. It is meant to be a precise way of giving information about the function without a rather lengthy written explanation.
- The most popular function notation is $f(x)$ which is read "f of x". Any letter can be used to name a function. for example: $g(x)$, $h(x)$, etc.
- To evaluate a function, plug in the input (the given value or expression) for the function's variable (place holder, x).

Examples:

1) Evaluate: $h(n) = n - 2$, find $h(2)$

Solution: Substitute n with 4: Then: $h(n) = n - 2 \rightarrow h(2) = (2)^2 - 2 \rightarrow h(2) = 4 - 2 = 2$

2) Evaluate: $w(x) = 5x - 1$, find $w(3)$.

Solution: Substitute x with 3: Then: $w(x) = 5x - 1 \rightarrow w(3) = 5(3) - 1 = 15 - 1 = 14$

3) Evaluate: $f(x) = x^2 - 2$, find $h(2)$.

Solution: Substitute x with 2: Then: $f(x) = x^2 - 2 \rightarrow f(2) = (2)^2 - 2 = 4 - 2 = 2$

4) Evaluate: $p(x) = 2x^2 - 4$, find $p(3a)$.

Solution: Substitute x with $3a$: Then: $p(x) = 2x^2 - 4 \rightarrow p(3a) = 2(3a)^2 - 4 \rightarrow$

$$p(3a) = 2(9a^2) - 4 = 18a^2 - 4$$

Adding and Subtracting Functions

- Just like we can add and subtract numbers and expressions, we can add or subtract two functions and simplify or evaluate them. The result is a new function.
- For two functions $f(x)$ and $g(x)$, we can create two new functions: $(f+g)(x) = f(x) + g(x)$ and $(f-g)(x) = f(x) - g(x)$

Examples:

1) $g(x) = a - 1, f(a) = a + 2$, Find: $(g+f)(a)$

Solution: $(g+f)(a) = g(a) + f(a)$

Then: $(g+f)(a) = (a-1) + (a+2) = 2a + 1$

2) $f(x) = 2x - 2, g(x) = x - 4$, Find: $(f-g)(x)$

Solution: $(f-g)(x) = f(x) - g(x)$

Then: $(f-g)(x) = (2x-2) - (x-4) = 2x - 2 - x + 4 = x + 2$

3) $g(x) = x^2 - 4, f(x) = 2x + 3$, Find: $(g+f)(x)$

Solution: $(g+f)(x) = g(x) + f(x)$

Then: $(g+f)(x) = (x^2 - 4) + (2x + 3) = x^2 + 2x - 1$

4) $f(x) = 2x^2 + 5, g(x) = 3x - 1$, Find: $(f-g)(5)$

Solution: $(f-g)(x) = f(x) - g(x)$

Then: $(f-g)(x) = (2x^2 + 5) - (3x - 1) = 2x^2 + 5 - 3x + 1 = 2x^2 - 3x + 6$

Substitute x with 5: $(g-f)(5) = 2(5)^2 - 3(5) + 6 = 50 - 15 + 6 = 41$

Multiplying and Dividing Functions

- Just like we can multiply and divide numbers and expressions, we can multiply and divide two functions and simplify or evaluate them.
- For two functions $f(x)$ and $g(x)$, we can create two new functions: $(f.g)(x) = f(x).g(x)$ and $\left(\frac{f}{g}\right)(x) = \frac{f(x)}{g(x)}$

Examples:

1) $g(x) = x - 2, f(x) = x + 3$, Find: $(g.f)(x)$

Solution: $(g.f)(x) = g(x).f(x) = (x-2)(x+3) = x^2 + 3x - 2x - 6$

2) $f(x) = x + 4, h(x) = x - 6$, Find: $\left(\frac{f}{h}\right)(x)$

Solution: $\left(\frac{f}{h}\right)(x) = \frac{f(x)}{h(x)} = \frac{x+4}{x-6}$

3) $g(x) = x + 5, f(x) = x - 2$, Find: $(g.f)(4)$

Solution: $(g.f)(x) = g(x).f(x) = (x+5)(x-2) = x^2 - 2x + 5x - 10 = x^2 + 3x - 10$

Substitute x with 4: $(g.f)(x) = (4)^2 + 3(4) - 10 = 16 + 12 - 10 = 18$

4) $f(x) = 2x + 3, h(x) = x + 8$, Find: $\left(\frac{f}{h}\right)(-1)$

Solution: $\left(\frac{f}{h}\right)(x) = \frac{f(x)}{h(x)} = \frac{2x+3}{x+8}$

Substitute x with -1: $\left(\frac{f}{h}\right)(x) = \frac{2x+3}{x+8} = \frac{2(-1)+3}{(-1)+8} = \frac{1}{7}$

Composition of Functions

- "Composition of functions" simply means combining two or more functions in a way where the output from one function becomes the input for the next function.
- The notation used for composition is: $(fog)(x) = f(g(x))$ and is read "f composed with g of x" or "f of g of x".

Examples:

1) *Using* $\mathrm{f(x) = x - 5}$ *and* $\mathrm{g(x) = 2x}$, *find:* $\mathrm{(fog)(x)}$

Solution: $(fog)(x) = f\big(g(x)\big)$. Then: $(fog)(x) = f\big(g(x)\big) = f(2x)$

Now find $f(2x)$ by substituting x with $2x$ in $f(x)$ function. Then: $\mathrm{f(x) = x - 5}$

$(x \to 2x) \to f(2x) = (2x) - 5 = 2x - 5$

2) *Using* $\mathrm{f(x) = x + 6}$ *and* $\mathrm{g(x) = x - 2}$, *find:* $\mathrm{(g\ o\ f)(-1)}$

Solution: $(f\ o\ g)(x) = f(g(x))$. *Then:* $(g\ o\ f)(x) = g\big(f(x)\big) = g(\mathrm{x} + 6)$, *now substitute* x *in* $\mathrm{g(x)}$ *by* $(\mathrm{x} + 6)$. *Then:* $g(\mathrm{x} + 6) = (x + 6) - 2 = x + 6 - 2 = x + 4$

Substitute x with -1: $(g\ o\ f)(-1) = g\big(f(x)\big) = x + 4 = -1 + 4 = 3$

3) *Using* $\mathrm{f(x) = 2x - 2}$ *and* $\mathrm{g(x) = 2x}$, *find:*$f(g(5))$

Solution: First find $g(5)$): $\mathrm{g(x) = 2x \to g(5) = 2(5) = 10}$

Then: $f\big(g(5)\big) = f(10)$. Now, find $f(10)$ by substituting x with 10 in $f(x)$ function.

Then: $f\big(g(5)\big) = f(10) = 2(10) - 2 = 20 - 2 = 18$

Chapter 13: Practices

Evaluate each function.

1) $g(n) = 6n - 3$, find $g(-2)$

______

2) $h(x) = -8x + 12$, find $h(3)$

______

3) $k(n) = 14 - 3n$, find $k(3)$

______

4) $g(x) = 4x - 4$, find $g(-2)$

______

5) $k(n) = 8n - 7$, find $k(4)$

______

6) $w(n) = -2n + 14$, find $w(5)$

______

Perform the indicated operation.

7) $f(x) = x + 6$
   $g(x) = 3x + 3$
   Find $(f - g)(2)$

______

8) $g(x) = x - 3$
   $f(x) = -x - 4$
   Find $(g - f)(-2)$

______

9) $h(t) = 5t + 4$
   $g(t) = 2t + 2$
   Find $(h + g)(-1)$

______

10) $g(a) = 3a - 5$
   $f(a) = a^2 + 6$
   Find $(g + f)(3)$

______

11) $g(x) = 4x - 5$
   $h(x) = 6x^2 + 5$
   Find $(g - f)(-2)$

______

12) $h(x) = x^2 + 3$
   $g(x) = -4x + 1$
   Find $(h + g)(4)$

______

Perform the indicated operation.

13) $g(x) = x + 2$
$f(x) = x + 3$
Find $(g.f)(4)$

_____

14) $f(x) = 2x$
$h(x) = -x + 6$
Find $(f.h)(-2)$

_____

15) $g(a) = a + 2$
$h(a) = 2a - 3$
Find $(g.h)(5)$

_____

16) $f(x) = 2x + 4$
$h(x) = 4x - 2$
Find $(\frac{f}{h})(2)$

_____

17) $f(x) = a^2 - 2$
$g(x) = -4 + 3a$
Find $(\frac{f}{g})(2)$

_____

18) $g(a) = 4a + 6$
$f(a) = 2a - 8$
Find $(\frac{g}{f})(3)$

_____

Using $f(x) = 2x + 5$ ***and*** $g(x) = x - 2$***, find:***

19) $g(f(2)) =$ ____

20) $g(f(-2)) =$ ____

21) $f(g(5)) =$ ____

22) $f(f(4)) =$ ____

23) $g(f(3)) =$ ____

24) $g(f(-3)) =$ ____

Answers – Chapter 13

1) -15
2) -12
3) 5
4) -12
5) 25
6) 4
7) -1
8) -3
9) -1
10) 19
11) -42
12) 4
13) 42
14) -32
15) 49
16) $\frac{4}{3}$
17) 1
18) -9
19) 7
20) -1
21) 11
22) 31
23) 9
24) -3

SSAT Upper Level Test Review

The SSAT, or Secondary School Admissions Test, is a standardized test to help determine admission to private elementary, middle and high schools.

There are currently three Levels of the SSAT:

- ✓ Lower Level (for students in 3rd and 4th grade)
- ✓ Middle Level (for students in 5th-7th grade)
- ✓ Upper Level (for students in 8th-11th grade)

There are six sections on the SSAT Upper Level Test:

- ✓ Writing: 25 minutes.
- ✓ Math section: 25 questions, 30 minutes
- ✓ Reading section: 40 questions, 40 minutes
- ✓ Verbal section: 60 questions, 30 minutes
- ✓ Math section: 25 questions, 30 minutes
- ✓ Experimental: 16 questions, 15 minutes.

In this book, there are 2 complete SSAT Upper Level Math Practice Tests. Take these tests to see what score you'll be able to receive on a real SSAT Upper Level test.

Good luck!

Time to Test

Time to refine your skill with a practice examination

Take a practice SSAT Upper Level Mathematics Test to simulate the test day experience. After you've finished, score your test using the answer keys.

Before You Start

- You'll need a pencil and a timer to take the test.
- Each test contains 25 multiple-choice questions. For each question, there are five possible answers. Choose which one is best.
- After you've finished the test, review the answer key to see where you went wrong.
- Use the answer sheet provided to record your answers. (You can cut it out or photocopy it)
- You will receive 1 point for every correct answer, and you will lose $\frac{1}{4}$ point for each incorrect answer. There is no penalty for skipping a question.

Calculators are NOT permitted for the SSAT Upper Level Test

Good Luck!

SSAT Upper Level Math

Practice Test 1

2020 - 2021

Two Parts

Total number of questions: 50

Section 1: 25 questions

Section 2: 25 questions

Total time for two parts: 60 Minutes

SSAT Upper Level Math Practice Test 1 Answer Sheet

Remove (or photocopy) this answer sheet and use it to complete the practice test.

SSAT Upper Level Mathematics Practice Test 1 Answer Sheet

SSAT Upper Level Practice Section 1

1	Ⓐ Ⓑ Ⓒ Ⓓ Ⓔ	11	Ⓐ Ⓑ Ⓒ Ⓓ Ⓔ	21	Ⓐ Ⓑ Ⓒ Ⓓ Ⓔ
2	Ⓐ Ⓑ Ⓒ Ⓓ Ⓔ	12	Ⓐ Ⓑ Ⓒ Ⓓ Ⓔ	22	Ⓐ Ⓑ Ⓒ Ⓓ Ⓔ
3	Ⓐ Ⓑ Ⓒ Ⓓ Ⓔ	13	Ⓐ Ⓑ Ⓒ Ⓓ Ⓔ	23	Ⓐ Ⓑ Ⓒ Ⓓ Ⓔ
4	Ⓐ Ⓑ Ⓒ Ⓓ Ⓔ	14	Ⓐ Ⓑ Ⓒ Ⓓ Ⓔ	24	Ⓐ Ⓑ Ⓒ Ⓓ Ⓔ
5	Ⓐ Ⓑ Ⓒ Ⓓ Ⓔ	15	Ⓐ Ⓑ Ⓒ Ⓓ Ⓔ	25	Ⓐ Ⓑ Ⓒ Ⓓ Ⓔ
6	Ⓐ Ⓑ Ⓒ Ⓓ Ⓔ	16	Ⓐ Ⓑ Ⓒ Ⓓ Ⓔ		
7	Ⓐ Ⓑ Ⓒ Ⓓ Ⓔ	17	Ⓐ Ⓑ Ⓒ Ⓓ Ⓔ		
8	Ⓐ Ⓑ Ⓒ Ⓓ Ⓔ	18	Ⓐ Ⓑ Ⓒ Ⓓ Ⓔ		
9	Ⓐ Ⓑ Ⓒ Ⓓ Ⓔ	19	Ⓐ Ⓑ Ⓒ Ⓓ Ⓔ		
10	Ⓐ Ⓑ Ⓒ Ⓓ Ⓔ	20	Ⓐ Ⓑ Ⓒ Ⓓ Ⓔ		

SSAT Upper Level Practice Section 2

1	Ⓐ Ⓑ Ⓒ Ⓓ Ⓔ	11	Ⓐ Ⓑ Ⓒ Ⓓ Ⓔ	21	Ⓐ Ⓑ Ⓒ Ⓓ Ⓔ
2	Ⓐ Ⓑ Ⓒ Ⓓ Ⓔ	12	Ⓐ Ⓑ Ⓒ Ⓓ Ⓔ	22	Ⓐ Ⓑ Ⓒ Ⓓ Ⓔ
3	Ⓐ Ⓑ Ⓒ Ⓓ Ⓔ	13	Ⓐ Ⓑ Ⓒ Ⓓ Ⓔ	23	Ⓐ Ⓑ Ⓒ Ⓓ Ⓔ
4	Ⓐ Ⓑ Ⓒ Ⓓ Ⓔ	14	Ⓐ Ⓑ Ⓒ Ⓓ Ⓔ	24	Ⓐ Ⓑ Ⓒ Ⓓ Ⓔ
5	Ⓐ Ⓑ Ⓒ Ⓓ Ⓔ	15	Ⓐ Ⓑ Ⓒ Ⓓ Ⓔ	25	Ⓐ Ⓑ Ⓒ Ⓓ Ⓔ
6	Ⓐ Ⓑ Ⓒ Ⓓ Ⓔ	16	Ⓐ Ⓑ Ⓒ Ⓓ Ⓔ		
7	Ⓐ Ⓑ Ⓒ Ⓓ Ⓔ	17	Ⓐ Ⓑ Ⓒ Ⓓ Ⓔ		
8	Ⓐ Ⓑ Ⓒ Ⓓ Ⓔ	18	Ⓐ Ⓑ Ⓒ Ⓓ Ⓔ		
9	Ⓐ Ⓑ Ⓒ Ⓓ Ⓔ	19	Ⓐ Ⓑ Ⓒ Ⓓ Ⓔ		
10	Ⓐ Ⓑ Ⓒ Ⓓ Ⓔ	20	Ⓐ Ⓑ Ⓒ Ⓓ Ⓔ		

SSAT Upper Level Math
Practice Test 1

Section 1

25 questions

Total time for this section: 30 Minutes

You may NOT use a calculator for this test.

1) A school wants to give each of its 22 top students a football ball. If the balls are in boxes of four, how many boxes of balls they need to purchase?
A. 2
B. 4
C. 6
D. 8
E. 22

2) A shaft rotates 240 times in 12 seconds. How many times does it rotate in 20 seconds?
A. 450
B. 400
C. 300
D. 250
E. 200

3) What is the value of x in the following figure?
A. 150
B. 145
C. 135
D. 125
E. 115

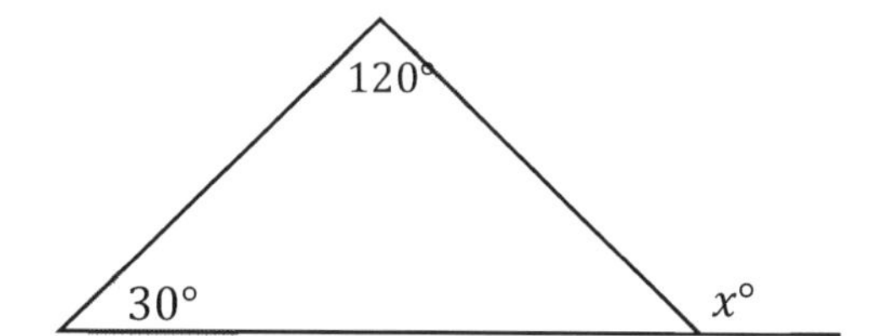

4) $14.125 \div 0.005$?
A. 2.825
B. 28.250
C. 282.50
D. 2825
E. 28025

5) How many tiles of 5 cm^2 is needed to cover a floor of dimension 5 cm by 20 cm?
A. 5
B. 10
C. 15
D. 20
E. 25

6) If $\frac{30}{A} + 1 = 7$, then $30 + A = ?$
A. 2
B. 7
C. 35
D. 40
E. 50

7) What is the value of the sum of the tens and thousandths in number 3,617.89652?

A. 16
B. 11
C. 8
D. 14
E. 6

8) Jack earns $720 for his first 45 hours of work in a week and is then paid 1.5 times his regular hourly rate for any additional hours. This week, Jack needs $936 to pay his rent, bills and other expenses. How many hours must he work to make enough money in this week?

A. 50
B. 54
C. 60
D. 63
E. 64

9) If 50% of a number is 85, then what is the 20% of that number?

A. 34
B. 40
C. 50
D. 60
E. 85

10) Which of the following statements is correct, according to the graph below?

A. Number of books sold in April was twice the number of books sold in July.
B. Number of books sold in July was less than half the number of books sold in May.
C. Number of books sold in June was half the number of books sold in April.
D. Number of books sold in July was equal to the number of books sold in April plus the number of books sold in June.
E. More books were sold in April than in July.

11) If $x \blacksquare y = \sqrt{x^2 + y}$, what is the value of $6 \blacksquare 28$?

A. $\sqrt{168}$
B. 10
C. 8
D. 6
E. 4

12) If $2 \le x < 6$, what is the minimum value of the following expression?

$$3x + 1$$

A. 9
B. 7
C. 4
D. 2
E. 1

13) What is the answer of $0.9 \div 0.015$?

A. $\frac{1}{60}$
B. $\frac{1}{6}$
C. 6
D. 60
E. 600

14) $\dfrac{1\frac{4}{3}+\frac{1}{4}}{2\frac{1}{2}-\frac{17}{8}}$ is approximately equal to.

A. 3.33
B. 3.6
C. 5.67
D. 6.88
E. 6.97

15) There are four equal tanks of water. If $\frac{2}{5}$ of a tank contains 300 liters of water, what is the capacity of the four tanks of water together?

A. $1{,}500$ liters
B. $2{,}000$ liters
C. $2{,}500$ liters
D. $3{,}000$ liters
E. $3{,}500$ liters

16) A cruise line ship left Port A and traveled 30 miles due west and then 40 miles due north. At this point, what is the shortest distance from the cruise to port A?

A. 50 miles
B. 55 miles
C. 60 miles
D. 70 miles
E. 110 miles

17) The average weight of 20 girls in a class is $55\ kg$ and the average weight of 42 boys in the same class is $82\ kg$. What is the average weight of all the 62 students in that class?

A. $70\ kg$
B. $72.20\ kg$
C. $73.29\ kg$
D. $74.44\ kg$
E. $75.20\ kg$

18) Two-kilograms apple and two-kilograms orange cost $\$28.4$. If one-kilogram apple costs $\$5.2$, how much does one-kilogram orange cost?

A. $\$9$
B. $\$6.5$
C. $\$6$
D. $\$5.5$
E. $\$5$

19) David's current age is 44 years, and Ava's current age is 4 years old. In how many years David's age will be 5 times Ava's age?

A. 4
B. 6
C. 8
D. 10
E. 14

20) Michelle and Alec can finish a job together in 50 minutes. If Michelle can do the job by herself in 2.5 hours, how many minutes does it take Alec to finish the job?

A. 60
B. 75
C. 80
D. 100
E. 150

21) The sum of six different negative integers is -80. If the smallest of these integers is -16, what is the largest possible value of one of the other five integers?

A. -16
B. -10
C. -8
D. -4
E. -1

22) What is the slope of a line that is perpendicular to the line $2x - 4y = 24$?

A. -2
B. $-\frac{1}{2}$
C. 6
D. 12
E. 14

23) The width of a box is one third of its length. The height of the box is one half of its width. If the length of the box is $24\ cm$, what is the volume of the box?

A. $91\ cm^3$
B. $172\ cm^3$
C. $254\ cm^3$
D. $768\ cm^3$
E. $2{,}990\ cm^3$

24) A football team won exactly 70% of the games it played during last session. Which of the following could be the total number of games the team played last season?

A. 59
B. 45
C. 72
D. 20
E. 11

25) The Jackson Library is ordering some bookshelves. If x is the number of bookshelves the library wants to order, which each costs $\$200$ and there is a one-time delivery charge of $\$600$, which of the following represents the total cost, in dollar, per bookshelf?

A. $\frac{200x+600}{x}$
B. $\frac{200x+600}{200}$
C. $200 + 600x$
D. $200x + 600$
E. $200x - 600$

IF YOU FINISH BEFORE TIME IS CALLED, YOU MAY CHECK YOUR WORK ON THIS SECTION ONLY. DO NOT TURN TO ANY OTHER SECTION IN THE TEST. **STOP**

SSAT Upper Level Math

Practice Test 1

Section 2

25 questions

Total time for this section: 30 Minutes

You may NOT use a calculator for this test.

1) There are 14 marbles in the bag A and 18 marbles in the bag B. If the sum of the marbles in both bags will be shared equally between two children, how many marbles bag A has less than the marbles that each child will receive?

A. 2
B. 3
C. 4
D. 5
E. 6

2) If Jason's mark is F more than Alex, and Jason's mark is 18, which of the following can be Alex's mark?

A. $18 - F$
B. $F - 18$
C. $\frac{F}{18}$
D. $18F$
E. $18 + F$

3) When number $81{,}602$ is divided by 240, the result is closest to?

A. 4
B. 30
C. 200
D. 300
E. 340

4) The price of a sofa is decreased by 50% to $\$530$. What was its original price?

A. $\$580$
B. $\$620$
C. $\$760$
D. $\$900$
E. $\$1{,}060$

5) If the perimeter of the following figure be 33, what is the value of x?

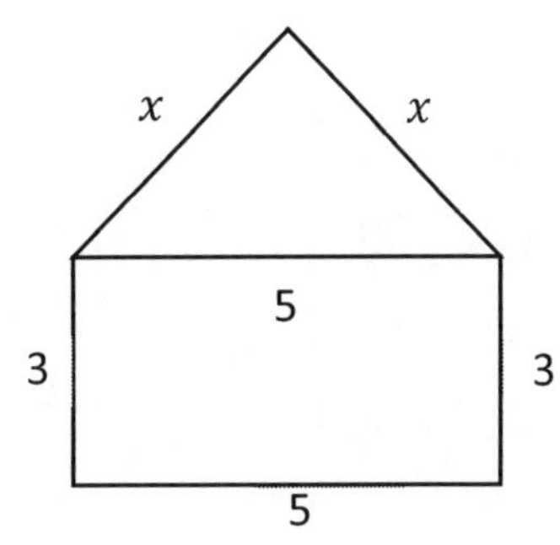

A. 3
B. 4
C. 6
D. 8
E. 11

6) To paint a wall with the area of $57m^2$, how many liters of paint do we need if each liter of paint is enough to paint a wall with dimension of $80\ cm \times 95\ cm$?

A. 50
B. 75
C. 100
D. 150
E. 200

7) Which of the following expression is not equal to 6?

A. $12 \times \frac{1}{2}$
B. $24 \times \frac{1}{4}$
C. $2 \times \frac{6}{2}$
D. $5 \times \frac{6}{5}$
E. $6 \times \frac{1}{6}$

8) $760 - 9\frac{8}{14} = ?$

A. $750\frac{3}{7}$
B. $750\frac{8}{7}$
C. $753\frac{1}{7}$
D. $753\frac{8}{7}$
E. $754\frac{1}{7}$

9) What is the missing term in the given sequence?

$$3, 4, 6, 9, 13, 18, 24, \_\_\_, 39$$

A. 25
B. 27
C. 28
D. 29
E. 31

10) A driver rests one hour and 15 minutes for every 3 hours driving. How many minutes will he rest if he drives 24 hours?

A. $3\ hours\ and\ 26\ minutest$
B. $4\ hours\ and\ 10\ minutest$
C. $5\ hours\ and\ 36\ minutest$
D. $6\ hours\ and\ 46\ minutest$
E. $10\ hours$

11) If $5y + 5 < 30$, then y could be equal to?

A. 14
B. 11
C. 9.5
D. 5
E. 3

12) If the area of the following rectangular $ABCD$ is 160, and E is the midpoint of AB, what is the area of the shaded part?

A. 30
B. 60
C. 70
D. 80
E. 90

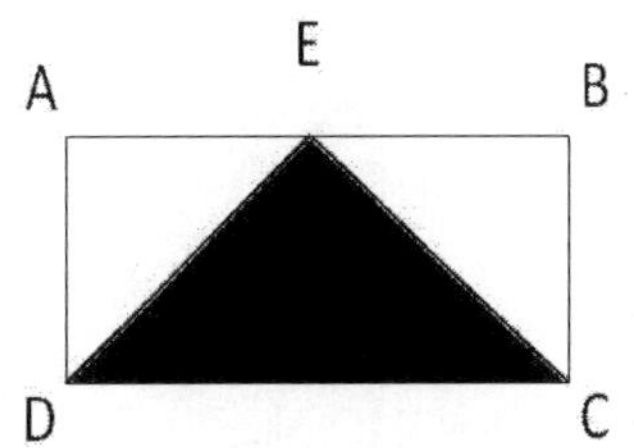

Questions 13 to 14 are based on the following graph

A library has 840 books that include Mathematics, Physics, Chemistry, English and History. Use following graph to answer questions 13 to 14.

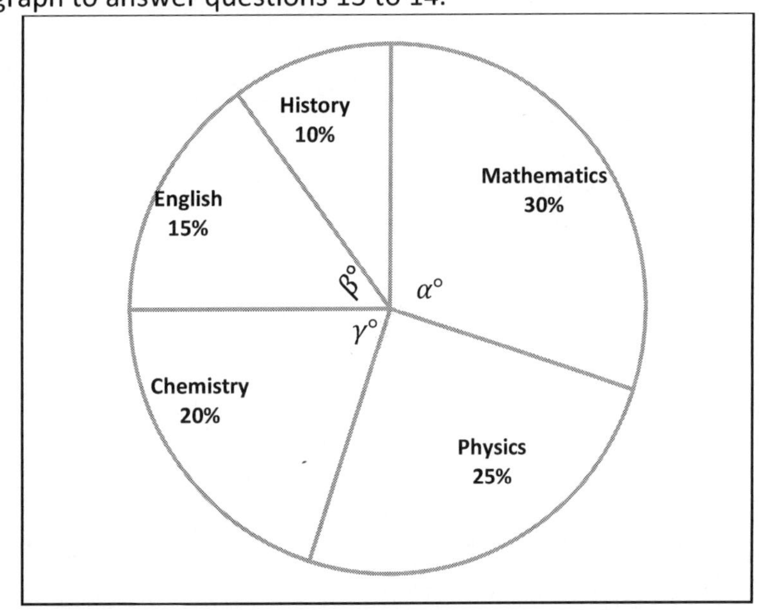

13) What is the product of the number of Mathematics and number of English books?

A. 21,168
B. 31,752
C. 26,460
D. 17,640
E. 35,280

14) What are the values of angle α and β respectively?

A. $90°, 54°$
B. $120°, 36°$
C. $120°, 45°$
D. $108°, 54°$
E. $108°, 45°$

15) The capacity of a red box is 20% bigger than the capacity of a blue box. If the red box can hold 60 equal sized books, how many of the same books can the blue box hold?

A. 15
B. 25
C. 40
D. 50
E. 60

16) If $8x + y = 24$ and $x - z = 17$, what is the value of x?

A. 1
B. 4
C. 9
D. 19
E. it cannot be determined from the information given

17) Find the perimeter of following shape.

A. 20
B. 21
C. 22
D. 30
E. 34

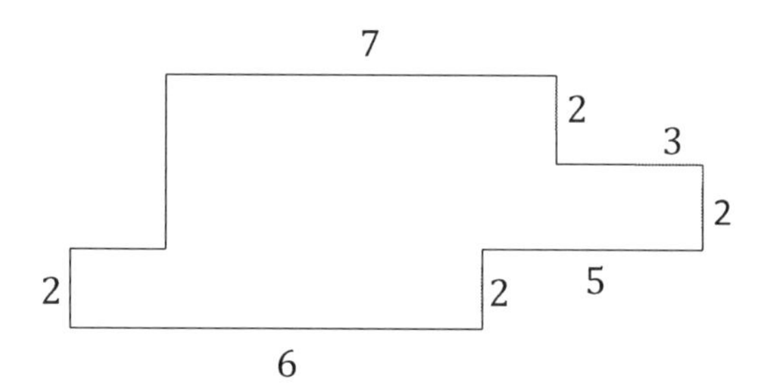

18) There are 60.4 liters of gas in a car fuel tank. In the first week and second week of April, the car uses 4.86 and 23.9 liters of gas respectively. If the car was park in the third week of April and 11.22 liters of gas will be added to the fuel tank, how many liters of gas are in the fuel tank of the car?

A. 20.41 liters
B. 26.5 liters
C. 28 liters
D. 28.79 liters
E. 42.86 liters

19) If $a \times b$ is divisible by 4, which of the following expression must also be divisible by 4?

A. $2a - 5b$
B. $3a - b$
C. $2a \times 3b$
D. $\frac{a}{b}$
E. $\frac{a \times b}{3}$

20) Which of the following could be the value of x if $\frac{6}{8} + x > 2$?

A. $\frac{1}{3}$
B. $\frac{3}{5}$
C. $\frac{6}{5}$
D. $\frac{4}{3}$
E. $\frac{2}{3}$

21) If a gas tank can hold 20 gallons, how many gallons does it contain when it is $\frac{3}{4}$ full?

A. 135
B. 52.5
C. 50
D. 20
E. 15

22) In the following figure, point Q lies on line n, what is the value of y if $x = 35$?

A. 10
B. 25
C. 30
D. 40
E. 50

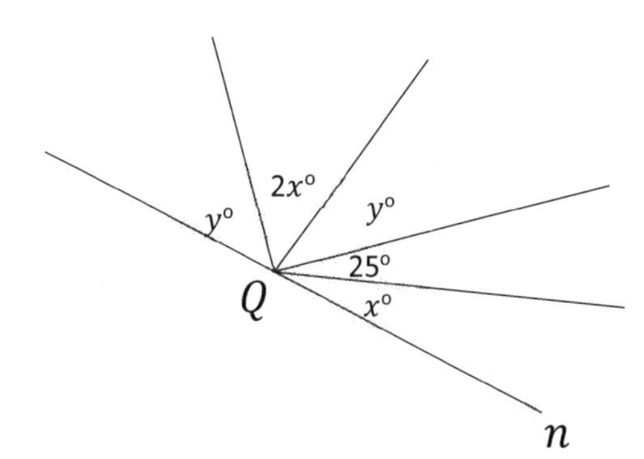

23) A number is chosen at random from 1 to 20. Find the probability of not selecting a composite number. (A composite number is a number that is divisible by itself, 1 and at least one other whole number)

A. $\frac{1}{20}$
B. $\frac{2}{5}$
C. $\frac{9}{20}$
D. 1
E. 0

24) $615 \div 4 = ?$

A. $\frac{600}{4} \times \frac{10}{4} \times \frac{5}{4}$
B. $600 + \frac{10}{4} + \frac{5}{4}$
C. $\frac{600}{4} + \frac{10}{4} + \frac{5}{4}$
D. $\frac{600}{4} \div \frac{10}{4} \div \frac{5}{4}$
E. $\frac{6}{4} + \frac{1}{4} + \frac{5}{4}$

25) What is the average of circumference of figure A and area of figure B? ($\pi = 3$)

A. 75
B. 70
C. 60
D. 50
E. 44

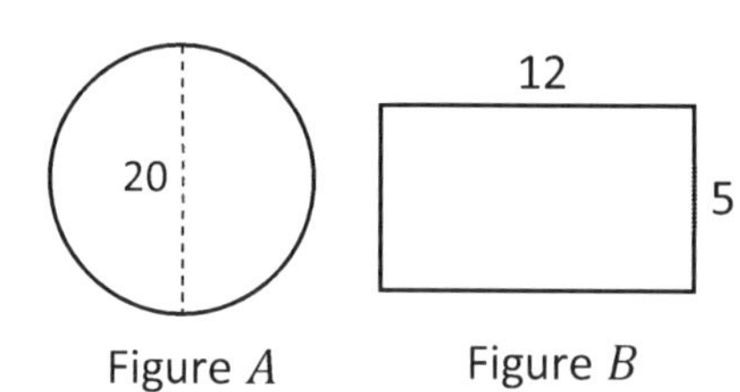

Figure A Figure B

IF YOU FINISH BEFORE TIME IS CALLED, YOU MAY CHECK YOUR WORK ON THIS SECTION ONLY. DO NOT TURN TO ANY OTHER SECTION IN THE TEST.

STOP

SSAT Upper Level Math

Practice Test 2

2020 - 2021

Two Parts

Total number of questions: 50

Section 1: 25 questions

Section 2: 25 questions

Total time for two parts: 60 Minutes

SSAT Upper Level Math Practice Test 2 Answer Sheet

Remove (or photocopy) this answer sheet and use it to complete the practice test.

SSAT Upper Level Mathematics Practice Test 2 Answer Sheet

SSAT Upper Level Math Section 1

1	Ⓐ Ⓑ Ⓒ Ⓓ Ⓔ	11	Ⓐ Ⓑ Ⓒ Ⓓ Ⓔ	21	Ⓐ Ⓑ Ⓒ Ⓓ Ⓔ
2	Ⓐ Ⓑ Ⓒ Ⓓ Ⓔ	12	Ⓐ Ⓑ Ⓒ Ⓓ Ⓔ	22	Ⓐ Ⓑ Ⓒ Ⓓ Ⓔ
3	Ⓐ Ⓑ Ⓒ Ⓓ Ⓔ	13	Ⓐ Ⓑ Ⓒ Ⓓ Ⓔ	23	Ⓐ Ⓑ Ⓒ Ⓓ Ⓔ
4	Ⓐ Ⓑ Ⓒ Ⓓ Ⓔ	14	Ⓐ Ⓑ Ⓒ Ⓓ Ⓔ	24	Ⓐ Ⓑ Ⓒ Ⓓ Ⓔ
5	Ⓐ Ⓑ Ⓒ Ⓓ Ⓔ	15	Ⓐ Ⓑ Ⓒ Ⓓ Ⓔ	25	Ⓐ Ⓑ Ⓒ Ⓓ Ⓔ
6	Ⓐ Ⓑ Ⓒ Ⓓ Ⓔ	16	Ⓐ Ⓑ Ⓒ Ⓓ Ⓔ		
7	Ⓐ Ⓑ Ⓒ Ⓓ Ⓔ	17	Ⓐ Ⓑ Ⓒ Ⓓ Ⓔ		
8	Ⓐ Ⓑ Ⓒ Ⓓ Ⓔ	18	Ⓐ Ⓑ Ⓒ Ⓓ Ⓔ		
9	Ⓐ Ⓑ Ⓒ Ⓓ Ⓔ	19	Ⓐ Ⓑ Ⓒ Ⓓ Ⓔ		
10	Ⓐ Ⓑ Ⓒ Ⓓ Ⓔ	20	Ⓐ Ⓑ Ⓒ Ⓓ Ⓔ		

SSAT Upper Level Math Section 2

1	Ⓐ Ⓑ Ⓒ Ⓓ Ⓔ	11	Ⓐ Ⓑ Ⓒ Ⓓ Ⓔ	21	Ⓐ Ⓑ Ⓒ Ⓓ Ⓔ
2	Ⓐ Ⓑ Ⓒ Ⓓ Ⓔ	12	Ⓐ Ⓑ Ⓒ Ⓓ Ⓔ	22	Ⓐ Ⓑ Ⓒ Ⓓ Ⓔ
3	Ⓐ Ⓑ Ⓒ Ⓓ Ⓔ	13	Ⓐ Ⓑ Ⓒ Ⓓ Ⓔ	23	Ⓐ Ⓑ Ⓒ Ⓓ Ⓔ
4	Ⓐ Ⓑ Ⓒ Ⓓ Ⓔ	14	Ⓐ Ⓑ Ⓒ Ⓓ Ⓔ	24	Ⓐ Ⓑ Ⓒ Ⓓ Ⓔ
5	Ⓐ Ⓑ Ⓒ Ⓓ Ⓔ	15	Ⓐ Ⓑ Ⓒ Ⓓ Ⓔ	25	Ⓐ Ⓑ Ⓒ Ⓓ Ⓔ
6	Ⓐ Ⓑ Ⓒ Ⓓ Ⓔ	16	Ⓐ Ⓑ Ⓒ Ⓓ Ⓔ		
7	Ⓐ Ⓑ Ⓒ Ⓓ Ⓔ	17	Ⓐ Ⓑ Ⓒ Ⓓ Ⓔ		
8	Ⓐ Ⓑ Ⓒ Ⓓ Ⓔ	18	Ⓐ Ⓑ Ⓒ Ⓓ Ⓔ		
9	Ⓐ Ⓑ Ⓒ Ⓓ Ⓔ	19	Ⓐ Ⓑ Ⓒ Ⓓ Ⓔ		
10	Ⓐ Ⓑ Ⓒ Ⓓ Ⓔ	20	Ⓐ Ⓑ Ⓒ Ⓓ Ⓔ		

SSAT Upper Level Math
Practice Test 2

Section 1

25 questions

Total time for this section: 30 Minutes

You may NOT use a calculator for this test.

1) What is the value of the "9" in number 131.493?
   A. 9 ones
   B. 9 tenths
   C. 9 hundredths
   D. 9 tens
   E. 9 thousandths

2) If $x - 20 = -20$, then $x \times 20 =$?
   A. 0
   B. 10
   C. 20
   D. 40
   E. 60

3) $0.04 \times 13.00 =?$
   A. 5.2
   B. 52.00
   C. 0.52
   D. 5.02
   E. 0.052

4) If Logan ran 3.5 miles in half an hour, his average speed was?
   A. 2.25 *miles per hour*
   B. 3.5 *miles per hour*
   C. 3.75 *miles per hour*
   D. 4.26 *miles per hour*
   E. 7 *miles per hour*

5) Given the diagram, what is the perimeter of the quadrilateral?

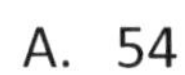

   A. 54
   B. 70
   C. 720
   D. 26740
   E. 55480

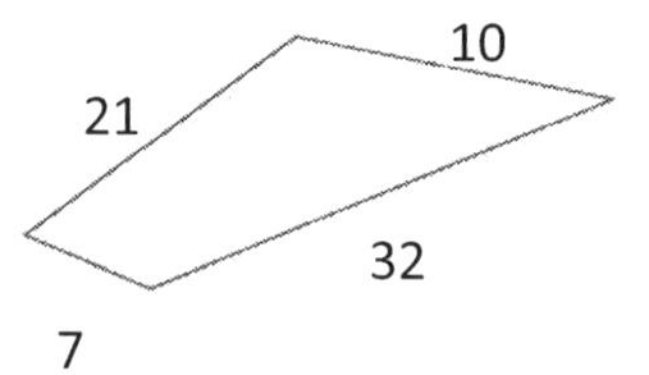

6) A pizza maker has M pounds of flour to make pizzas. After he has used 75 pounds of flour, how much flour is left? The expression that correctly represents the quantity of flour left is:
   A. $75 + M$
   B. $\frac{75}{M}$
   C. $75 - M$
   D. $M - 75$
   E. $75M$

7) Mia plans to buy a bracelet for every one of her 18 friends for their party. There are four bracelets in each pack. How many packs must she buy?
   A. 2
   B. 3
   C. 4
   D. 5
   E. 10

8) The distance between cities A and B is approximately $3{,}600$ miles. If Alice drives an average of 70 miles per hour, how many hours will it take Alice to drive from city A to city B?
   A. $Approximately\ 70\ hours$
   B. $Approximately\ 68\ hours$
   C. $Approximately\ 51 hours$
   D. $Approximately\ 37\ hours$
   E. $Approximately\ 21\ hours$

9) In a classroom of 60 students, 30 are female. What percentage of the class is male?
   A. 54%
   B. 50%
   C. 30%
   D. 26%
   E. 10%

10) A steak dinner at a restaurant costs $\$9.20$. If a man buys a steak dinner for himself and 4 friends, what will the total cost be?
   A. $\$33$
   B. $\$46$
   C. $\$28$
   D. $\$22.5$
   E. 12

11) An employee's rating on performance appraisals for the last three quarters were $93, 40$ and 87. If the required yearly average to qualify for the promotion is 97, what rating should the fourth quarter be?
   A. 158
   B. 168
   C. 178
   D. 188
   E. 198

12) Two third of 15 is equal to $\frac{2}{5}$ of what number?

A. 15
B. 25
C. 35
D. 45
E. 55

13) A cruise line ship left Port A and traveled 60 miles due west and then 80 miles due north. At this point, what is the shortest distance from the cruise to port A?

A. $100\ miles$
B. $120\ miles$
C. $130\ miles$
D. $150\ miles$
E. $170\ miles$

14) Last week $25{,}000$ fans attended a football match. This week three times as many bought tickets, but one sixth of them cancelled their tickets. How many are attending this week?

A. $49{,}000$
B. $52{,}200$
C. $62{,}500$
D. $72{,}500$
E. $84{,}700$

15) What is the slope of the line that is perpendicular to the line with equation $8x + y = 14$?

A. $\frac{1}{8}$
B. $-\frac{1}{8}$
C. $\frac{8}{12}$
D. 8
E. -8

16) In 1989, the average worker's income increased $\$3{,}000$ per year starting from $\$34{,}000$ annual salary. Which equation represents income greater than average? (I = income, x = number of years after 1989)

A. $I > 3{,}000\,x + 34{,}000$
B. $I > -\,3{,}000\,x + 34{,}000$
C. $I < -3{,}000\,x + 34{,}000$
D. $I < 3{,}000\,x - 34{,}000$
E. $I < 34{,}000\,x + 34{,}000$

17) In the following figure, MN is $50\ cm$. How long is ON?

A. $35\ cm$
B. $30\ cm$
C. $25\ cm$
D. $15\ cm$
E. $5\ cm$

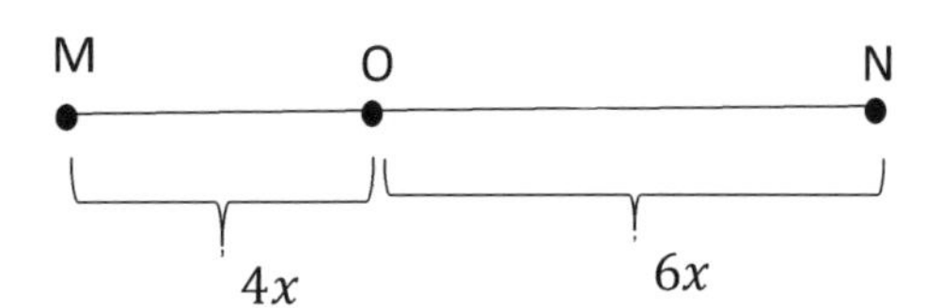

Questions 18 to 20 are based on the following data

The result of a research shows the number of men and women in four cities of a country.

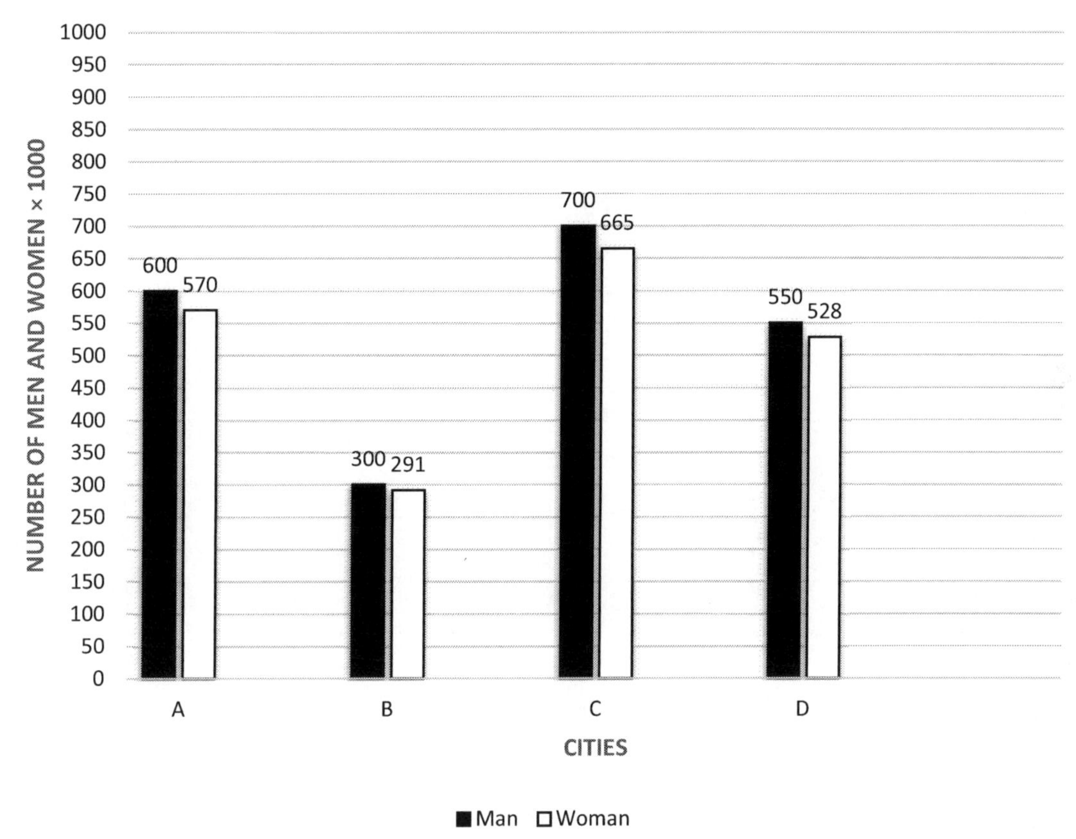

18) How many women should be added to city D until the ratio of women to men will be 1.2?

A. 120
B. 128
C. 132
D. 160
E. 165

19) What's the maximum ratio of woman to man in the four cities?

A. 0.98
B. 0.97
C. 0.96
D. 0.95
E. 0.93

20) What's the ratio of percentage of men in city A to percentage of women in city C?

A. 0.9
B. 0.95
C. 1
D. 1.05
E. 1.5

21) What is the difference of smallest 5–digit number and biggest 5–digit number?

A. $66,666$
B. $6,7899$
C. $8,8888$
D. $8,9999$
E. $9,9999$

22) Solve the following equation for y?

$$\frac{x}{3+4} = \frac{y}{11-8}$$

A. $\frac{3}{5}x$
B. $\frac{3}{7}x$
C. $3x$
D. x
E. $\frac{1}{7}x$

23) $\sqrt[5]{x^{21}} = ?$

A. $x^4\sqrt[5]{x}$
B. $20x$
C. x^{11}
D. x^{80}
E. x^4

24) and a width of 4 cm, what is the percent of ratio of the perimeter of rectangle B to rectangle A?

A. 10%
B. 20%
C. 62%
D. 75%
E. 143%

25) John traveled 160 km in 4 hours and Alice traveled 190 km in 5 hours. What is the ratio of the average speed of John to average speed of Alice?

A. 2 : 3
B. 3 : 2
C. 9 : 5
D. 6 : 5
E. 20 : 19

IF YOU FINISH BEFORE TIME IS CALLED, YOU MAY CHECK YOUR WORK ON THIS SECTION ONLY. DO NOT TURN TO ANY OTHER SECTION IN THE TEST. **STOP**

SSAT Upper Level Math

Practice Test 2

Section 2

25 questions

Total time for this section: 30 Minutes

You may NOT use a calculator for this test.

1) $0.44 \times 12.8 = ?$

A. 4.632
B. 4.965
C. 5.632
D. 5.891
E. 6.695

2) $4\frac{3}{8} \times 5\frac{1}{9} = ?$

A. $22\frac{1}{36}$
B. $22\frac{13}{36}$
C. $23\frac{4}{36}$
D. $23\frac{13}{36}$
E. 23

3) Which of the following is a whole number ?

A. $\frac{3}{2} \times \frac{5}{9}$
B. $\frac{1}{3} + \frac{1}{4}$
C. $\frac{40}{6}$
D. $3.5 + 2$
E. $3.5 + \frac{4}{8}$

4) 8 cubed is the same as:

A. 8×8
B. $8 \times 8 \times 8 \times 8$
C. 64
D. 512
E. $15{,}807$

5) If $\frac{2}{3}$ of a number equal to 16 then $\frac{5}{3}$ of the same number is:

A. 40
B. 35
C. 20
D. 14
E. 4

6) A swimming pool holds 3,000 cubic feet of water. The swimming pool is 30 feet long and 10 feet wide. How deep is the swimming pool?
A. 3 feet
B. 5 feet
C. 7 feet
D. 10 feet
E. 12 feet

7) We can put 25 colored pencils in each box and we have 450 colored pencils. How many boxes do we need?
A. 12
B. 15
C. 17
D. 18
E. 19

8) In the figure below, line A is parallel to line B. What is the value of angle x?
A. 45 degree
B. 55 degree
C. 80 degree
D. 120 degree
E. 145 degree

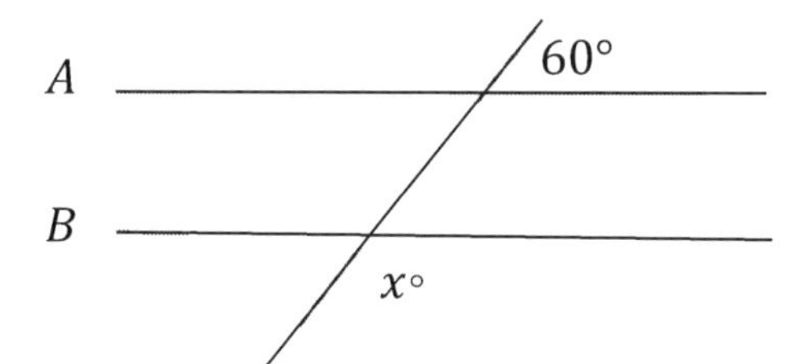

9) A bank is offering 4.5% simple interest on a savings account. If you deposit $13,500, how much interest will you earn in two years?
A. $420
B. $750
C. $6,300
D. $9,400
E. $1,215

10) Sophia purchased a sofa for $504. The sofa is regularly priced at $600. What was the percent discount Sophia received on the sofa?
A. 10%
B. 12%
C. 16%
D. 25%
E. 30%

11) How long does a 527–miles trip take moving at 62 miles per hour (mph)?

A. $7\ hours$
B. $7\ hours\ and\ 24\ minutes$
C. $8\ hours\ and\ 24\ minutes$
D. $8\ hours\ and\ 30\ minutes$
E. $9\ hours\ and\ 30\ minutes$

12) A construction company is building a wall. The company can build $40\ cm$ of the wall per minute. After 50 minutes construction, $\frac{2}{3}$ of the wall is completed. How high is the wall?

A. $10\ m$
B. $15\ m$
C. $30\ m$
D. $35\ m$
E. $42\ m$

13) When a number is subtracted from 28 and the difference is divided by that number, the result is 3. What is the value of the number?

A. 2
B. 5
C. 7
D. 12
E. 28

14) If car A drives 700 miles in 7 hours and car B drives the same distance in 5 hours, how many miles per hour does car B drive faster than car A?

A. 88 miles per hour
B. 65 miles per hour
C. 40 miles per hour
D. 12 miles per hour
E. 5 miles per hour

15) A company pays its employee \$8,000 plus 4% of all sales profit. If x is the number of all sales profit, which of the following represents the employee's revenue?

A. $0.04x$
B. $0.88x - 8{,}000$
C. $0.04x + 8{,}000$
D. $0.88x + 8{,}000$
E. $0.88x$

16) Which of the following shows the numbers in decreasing order?

A. $\frac{1}{5}, \frac{5}{3}, \frac{8}{11}, \frac{2}{3}$
B. $\frac{5}{3}, \frac{8}{11}, \frac{2}{3}, \frac{1}{5}$
C. $\frac{8}{11}, \frac{2}{3}, \frac{5}{3}, \frac{1}{5}$
D. $\frac{2}{3}, \frac{8}{11}, \frac{5}{3}, \frac{1}{5}$
E. None of the above

17) The ratio of boys to girls in a school is $3:2$. If there are 700 students in a school, how many boys are in the school?

A. 550
B. 500
C. 460
D. 420
E. 280

18) $\frac{(8+6)^2}{2} + 6 = ?$

A. 51
B. 80
C. 90
D. 100
E. 104

19) If $y = 5ab + 2b^4$ what is y when $a = 3$ and $= 2$?

A. 34
B. 35
C. 56
D. 62
E. 105

20) The area of a circle is 49π. What is the circumference of the circle?

A. 8π
B. 14π
C. 16π
D. 49π
E. 64π

21) If 80% of x equal to 40% of 20, then what is the value of $(x+7)^2$?

A. 27.27
B. 27
C. 29.01
D. 289
E. 12,026

22) What is the greatest common factor of 18 and 24?

A. 24
B. 18
C. 10
D. 6
E. 4

23) If the interior angles of a quadrilateral are in the ratio $1:2:3:4$, what is the measure of the smallest angle?

A. 36°
B. 72°
C. 108°
D. 144°
E. 180°

24) The length of a rectangle is $\frac{4}{5}$ times its width. If the width is 30, what is the perimeter of this rectangle?

A. 26
B. 58
C. 72
D. 92
E. 108

25) Find $\frac{2}{3}$ of $\frac{3}{4}$ of 160?

A. 80
B. 40
C. 20
D. 4
E. 2

IF YOU FINISH BEFORE TIME IS CALLED, YOU MAY CHECK YOUR WORK ON THIS SECTION ONLY. DO NOT TURN TO ANY OTHER SECTION IN THE TEST. **STOP**

SSAT Upper Level Math Practice Test Answer Keys

Now, it's time to review your results to see where you went wrong and what areas you need to improve.

SSAT Upper Level Math Practice Test 1								SSAT Upper Level Math Practice Test 2							
Section 1				Section 2				Section 1				Section 2			
1	C	16	A	1	A	16	E	1	C	16	A	1	C	16	B
2	B	17	C	2	A	17	E	2	A	17	B	2	B	17	D
3	A	18	A	3	E	18	E	3	C	18	C	3	E	18	E
4	D	19	B	4	E	19	C	4	E	19	B	4	D	19	D
5	D	20	B	5	E	20	D	5	B	20	D	5	A	20	B
6	C	21	B	6	B	21	E	6	D	21	D	6	D	21	D
7	D	22	A	7	E	22	B	7	D	22	B	7	D	22	D
8	B	23	D	8	A	23	B	8	C	23	A	8	D	23	A
9	A	24	D	9	E	24	C	9	B	24	C	9	E	24	E
10	C	25	A	10	E	25	C	10	B	25	E	10	C	25	A
11	C			11	E			11	B			11	D		
12	B			12	D			12	B			12	C		
13	D			13	B			13	A			13	C		
14	D			14	D			14	C			14	C		
15	D			15	E			15	A			15	C		

SSAT Upper Level Math Practice Test Answers and Explanations

SSAT Upper Level Mathematics Practice Test 1 Section 1

1) Choice C is correct

Number of packs equal to: $\frac{22}{4} \cong 5.5$, therefore, the school must purchase 6 packs.

2) Choice B is correct

First create a proportion: $\frac{240}{12} = \frac{x}{20}$, Number of rotates in 20 second equals to: $\frac{240 \times 20}{12} = 400$

3) Choice A is correct

$x = 30 + 120 = 150$

4) Choice D is correct

$14.125 \div 0.005 = \frac{\frac{14,125}{1,000}}{\frac{5}{1,000}} = \frac{14,125}{5} = 2,825$

5) Choice D is correct

The area of the floor is: $5\ cm \times 20\ cm = 100\ cm^2$, the number of tiles needed =

$$100 \div 5 = 20$$

6) Choice C is correct

$\frac{30}{\text{A}} + 1 = 7 \rightarrow \frac{30}{\text{A}} = 7 - 1 = 6 \rightarrow 30 = 6\text{A} \rightarrow \text{A} = \frac{30}{6} = 5, 30 + \text{A} = 30 + 5 = 35$

7) Choice D is correct

The digit in tens place is 8. The digit in the thousandths place is 6. Therefore; $8 + 6 = 14$

8) Choice B is correct

The amount of money that Jack earns for one hour: $\frac{\$720}{45} = \$16$

Number of additional hours that he needs to work in order to make enough money is: $\frac{\$936 - \$720}{1.5 \times \$16} = 9$, number of total hours Jack must work is: $45 + 9 = 54$

9) Choice A is correct

First, find the number. Let x be the number. Write the equation and solve for x.

50% of a number is 85, then: $0.5 \times x = 85 \rightarrow x = 85 \div 0.5 = 170$

20% of 170 is: $0.2 \times 170 = 34$

10) Choice C is correct

A. Number of books sold in April is: 380

Number of books sold in July is: $760 \rightarrow \frac{380}{760} = \frac{38}{76} = \frac{1}{2}$

B. Number of books sold in July is: 760

Half the number of books sold in May is: $\frac{1140}{2} = 570 \rightarrow 760 > 570$

C. Number of books sold in June is: 190

Half the number of books sold in April is: $\frac{380}{2} = 190 \rightarrow 190 = 190$

D. $380 + 190 = 570 < 760$

E. $380 < 760$

11) Choice C is correct

$6 \blacksquare 28 = \sqrt{6^2 + 28 =} \sqrt{36 + 28} = \sqrt{64} = 8$

12) Choice B is correct

$2 \leq x < 6 \rightarrow$ Multiply all sides of the inequality by 3. Then:

$2 \times 3 \leq 3 \times x < 6 \times 3 \rightarrow 6 \leq 3x < 18$

Add 1 to all sides. Then: $\rightarrow 6 + 1 \leq 3x + 1 < 18 + 1 \rightarrow \ 7 \leq 3x + 1 < 19$

Minimum value of $3x + 1$ is 7.

13) Choice D is correct

$0.90 \div 0.015 = \frac{0.90}{0.015} = \frac{\frac{90}{100}}{\frac{15}{1,000}} = \frac{90 \times 1,000}{15 \times 100} = \frac{90}{15} \times \frac{1,000}{100} = 6 \times 10 = 60$

14) Choice D is correct

$\frac{1\frac{4}{3}+\frac{1}{4}}{2\frac{1}{2}-\frac{17}{8}} = \frac{\frac{7}{3}+\frac{1}{4}}{\frac{5}{2}-\frac{17}{8}} = \frac{\frac{28+3}{12}}{\frac{20-1}{8}} = \frac{\frac{31}{12}}{\frac{3}{8}} = \frac{31 \times 8}{12 \times 3} = \frac{31 \times 2}{3 \times 3} = \frac{62}{9} \cong 6.88$

15) Choice D is correct

Let x be the capacity of one tank. Then, $\frac{2}{5}x = 300 \rightarrow x = \frac{300 \times 5}{2} = 750$ Liters

The amount of water in four tanks is equal to: $4 \times 750 = 3,000$ Liters

16) Choice A is correct

Use the information provided in the question to draw the shape.

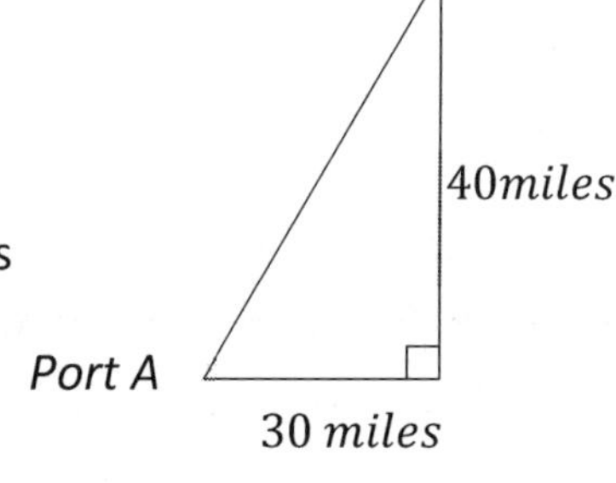

Use Pythagorean Theorem: $a^2 + b^2 = c^2$

$30^2 + 40^2 = c^2 \Rightarrow 900 + 1,600 = c^2 \Rightarrow 2500 = c^2 \Rightarrow c = 50$ miles

17) Choice C is correct

$Average = \frac{sum\ of\ terms}{number\ of\ terms}$,

The sum of the weight of all girls is: $20 \times 55 = 1,100\ kg$, The sum of the weight of all boys is: $42 \times 82 = 3,444\ kg$, The sum of the weight of all students is: $1,100 + 3,444 = 4,544 kg$

The average weight of the 62 students: $\frac{4,544}{62} = 73.29\ kg$

18) Choice A is correct

Let x be the cost of one-kilogram orange, then: $2x + (2 \times 5.2) = 28.4 \rightarrow$

$$2x + 10.4 = 28.4 \rightarrow 2x = 28.4 - 10.4 \rightarrow 2x = 18 \rightarrow x = \frac{18}{2} = \$9$$

19) Choice B is correct

Let's review the choices provided.

A. 4. In 4 years, David will be 48 and Ava will be 8.48 is not 5 times 8.
B. 6. In 6 years, David will be 50 and Ava will be 10. 50 is 5 times 10!
C. 8. In 8 years, David will be 52 and Ava will be 12. 52 is not 5 times 12.
D. 10. In 10 years, David will be 54 and Ava will be 14. 54 is not 5 times 14.
E.14. In 14 years, David will be 58 and Ava will be 18.58 is not 5 times 18.

20) Choice B is correct

Let b be the amount of time Alec can do the job, (change 2.5 hours to 150 minutes) then:

$\frac{1}{a} + \frac{1}{b} = \frac{1}{50} \rightarrow \frac{1}{150} + \frac{1}{b} = \frac{1}{50} \rightarrow \frac{1}{b} = \frac{1}{50} - \frac{1}{150} = \frac{2}{150} = \frac{1}{75}$, Then: $b = 75$ minutes

21) Choice B is correct

The smallest number is -16. To find the largest possible value of one of the other five integers, we need to choose the smallest possible integers for four of them. Let x be the largest number. Then: $-80 = (-16) + (-15) + (-14) + (-13) + (-12) + x \rightarrow$

$-80 = -70 + x, \rightarrow x = -80 + 70 = -10$

22) Choice A is correct

The equation of a line in slope intercept form is: $y = mx + b$, Solve for y.

$$2x - 4y = 24 \Rightarrow -4y = 24 - 2x \Rightarrow y = (24 - 2x) \div (-4) \Rightarrow y = \frac{1}{2}x - 6$$

The slope is $\frac{1}{2}$. The slope of the line perpendicular to this line is:

$$m_1 \times m_2 = -1 \Rightarrow \frac{1}{2} \times m_2 = -1 \Rightarrow m_2 = -2$$

23) Choice D is correct

If the length of the box is 24, then the width of the box is one third of it, 8, and the height of the box is 4 (half of the width). The volume of the box is:

$V = (length)(width)(height) = (24)(8)(4) = 768\ cm^3$

24) Choice D is correct

Choices A, B, C and E are incorrect because 70% of each of the numbers is non-whole number.

A. 59, 70% of $59 = 0.70 \times 59 = 41.3$
B. 45, 70% of $45 = 0.70 \times 45 = 31.5$
C. 72, 70% of $72 = 0.70 \times 72 = 50.4$
D. 20, 70% of $20 = 0.70 \times 20 = 14$
E. 11, 70% of $11 = 0.70 \times 11 = 7.7$

25) Choice A is correct

The amount of money for x bookshelf is: $200x$, Then, the total cost of all bookshelves is equal to: $200x + 600$, The total cost, in dollar, per bookshelf is: $\frac{Total\ cost}{number\ of\ items} = \frac{200x+600}{x}$

SSAT Upper Level Mathematics Practice Test 1 Section 2

1) Choice A is correct

$\frac{14+}{2} = \frac{32}{2} = 16$ Then, $16 - 14 = 2$

2) Choice A is correct

Alex's mark is F less than Jason's mark. Then, from the choices provided Alex's mark can only be $18 - F$.

3) Choice E is correct

$\frac{81602}{240} \cong 340.0083 \cong 340$

4) Choice E is correct

Let x be the original price. If the price of the sofa is decreased by 50% to \$530, then:

50% of $x = 530 \Rightarrow 0.50x = 530 \Rightarrow x = 530 \div 0.50 = 1{,}060$

5) Choice E is correct

Let's review the choices provided:

A. $x = 3 \rightarrow$ The perimeter of the figure is: $3 + 5 + 3 + 3 + 3 = 17 \neq 33$
B. $x = 4 \rightarrow$ The perimeter of the figure is: $3 + 5 + 3 + 4 + 4 = 19 \neq 33$
C. $x = 6 \rightarrow$ The perimeter of the figure is: $3 + 5 + 3 + 6 + 6 = 23 \neq 33$
D. $x = 8 \rightarrow$ The perimeter of the figure is: $3 + 5 + 3 + 8 + 8 = 27 \neq 33$
E. $x = 11 \rightarrow$ The perimeter of the figure is: $3 + 5 + 3 + 11 + 11 = 33 = 33$

6) Choice B is correct

The Area that one liter of paint is required: $80cm \times 95cm = 7{,}600cm^2$

Remember: $1\ m^2 = 10{,}000\ cm^2$ $(100 \times 100 = 10{,}000), then, 7{,}600cm^2 = 0.76\ m^2$

Number of liters of paint we need: $\frac{57}{0.76} = 75$ liters

7) Choice E is correct

Let's review the choices provided:

A. $12 \times \frac{1}{2} = \frac{12}{2} = 6 = 6$
B. $24 \times \frac{1}{4} = \frac{24}{4} = 6 = 6$
C. $2 \times \frac{6}{2} = \frac{12}{2} = 6 = 6$
D. $5 \times \frac{6}{5} = \frac{30}{5} = 6 = 6$
E. $6 \times \frac{1}{6} = \frac{6}{6} = 1 \neq 6$

8) Choice A is correct

$760 - 9\frac{8}{14} = (759 - 9) + \left(\frac{14}{14} - \frac{8}{14}\right) = 750\frac{3}{7}$

9) Choice E is correct

Find the difference of each pairs of numbers: $3, 4, 6, 9, 13, 18, 24, \_\_, 39$

The difference of 3 and 4 is 1, 4 and 6 is 2, 6 and 9 is 3, 9 and 13 is 4, 13 and 18 is 5, 18 and 24 is 6, 24 and next number should be 7. The number is $24 + 7 = 31$

10) Choice E is correct

Number of times that the driver rests $= \frac{24}{3} = 8$, Driver's rest time $=$ $1\ hour\ and\ 15\ minutes = 75\ minutes$, Then, $8 \times 75\ minutes = 600\ minutes$

$1\ hour = 60\ minutes = 600\ minutes = 10\ hours$

11) Choice E is correct

$5y + 5 < 30 \rightarrow 5y < 30 - 5 \rightarrow 5y < 25 \rightarrow y < 5$, the only choice that is less than 5 is E.

12) Choice D is correct

Since, E is the midpoint of AB, then the area of all triangles DAE, DEF, CFE and CBE are equal. Let x be the area of one of the triangle, then: $4x = 160 \rightarrow x = 40$

The area of $DEC = 2x = 2(40) = 80$

13) Choice B is correct

Number of Mathematics book: $0.3 \times 840 = 252$, Number of English books: $0.15 \times 840 = 126$, Product of number of Mathematics and number of English books: $252 \times 126 = 31{,}752$

14) Choice D is correct

The angle α is: $0.3 \times 360 = 108°$, The angle β is: $0.15 \times 360 = 54°$

15) Choice E is correct

The capacity of a red box is 20% bigger than the capacity of a blue box and it can hold 60 books. Therefore, we want to find a number that 20% bigger than that number is 60. Let x be that number. Then: $1.20 \times x = 60$, Divide both sides of the equation by 1.2. Then:

$x = \frac{60}{1.20} = 50$

16) Choice E is correct

We have two equations and three unknown variables, therefore x cannot be obtained.

17) Choice E is correct

$x + 2 = 2 + 2 + 2 \rightarrow x = 4$

$y + 7 + 3 = 6 + 5 \rightarrow y + 10 = 11 \rightarrow y = 1$

Then, the perimeter is:

$2 + 6 + 2 + 5 + 2 + 3 + 2 + 7 + 4 + 1 = 34$

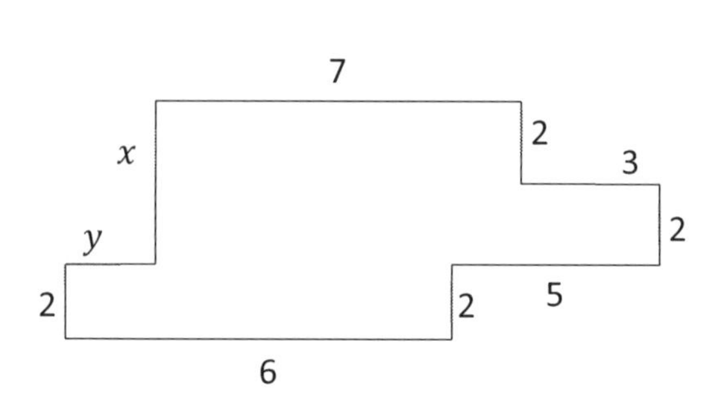

18) Choice E is correct

Amount of available petrol in tank: $60.4 - 4.86 - 23.9 + 11.22 = 42.86$ liters

19) Choice C is correct

Let put some values for a and b. If $a = 8$ and $b = 2$ $\rightarrow a \times b = 16$ $\rightarrow \frac{16}{4} = 8 \rightarrow 16$ is divisible by 4 then;

A. $2a - 5b = (2 \times 8) - (5 \times 2) = 6$ is not divisible by 4
B. $3a - b = (3 \times 8) - 2 = 24 - 2 = 22$ is not divisible by 4

If $a = 10$ and $b = 4$ $\rightarrow a \times b = 40$ $\rightarrow \frac{40}{4} = 10$ is not divisible by 4 then;

C. $2a \times 3b = 20 \times (3 \times 4) = 20 \times 12 = 240$ is divisible by 4
D. $\frac{a}{b} = \frac{10}{4}$ is not divisible by 4
E. $\frac{a \times b}{3} = \frac{10 \times 4}{3} = 13.33$

For choice C, if you try any other numbers for a and b, you will get the same result.

20) Choice D is correct

Let's review the choices provided:

A. $x = \frac{1}{3} \rightarrow \frac{6}{8} + \frac{1}{3} = \frac{18+8}{24} = \frac{26}{24} \cong 1.083 < 2$
B. $x = \frac{3}{5} \rightarrow \frac{6}{8} + \frac{3}{5} = \frac{30+2}{40} = \frac{54}{40} \cong 1.35 < 2$
C. $x = \frac{6}{5} \rightarrow \frac{6}{8} + \frac{6}{5} = \frac{30+48}{40} = \frac{78}{40} \cong 1.95 < 2$
D. $x = \frac{4}{3} \rightarrow \frac{6}{8} + \frac{4}{3} = \frac{18+}{24} = \frac{50}{24} \cong 2.08 > 2$
E. $x = \frac{2}{3} \rightarrow \frac{6}{8} + \frac{2}{3} = \frac{18+1}{24} = \frac{34}{24} \cong 1.41 < 2$

Only choice D is correct.

21) Choice E is correct

$\frac{3}{4} \times 20 = \frac{60}{4} = 15$

22) Choice B is correct

The angles on a straight line add up to 180 degrees. Let's review the choices provided:

A. $y = 10 \rightarrow x + 25 + y + 2x + y = 35 + 25 + 10 + 2(35) + 10 = 150 \neq 180$
B. $y = 25 \rightarrow x + 25 + y + 2x + y = 35 + 25 + 25 + 2(35) + 25 = 180$
C. $y = 30 \rightarrow x + 25 + y + 2x + y = 35 + 25 + 30 + 2(35) + 30 = 190 \neq 180$
D. $y = 40 \rightarrow x + 25 + y + 2x + y = 35 + 25 + 40 + 2(35) + 40 = 210 \neq 180$
E. $y = 50 \rightarrow x + 25 + y + 2x + y = 35 + 25 + 50 + 2(35) + 50 = 230 \neq 180$

23) Choice B is correct

Set of numbers that are not composite between 1 and 20: $A = \{2, 3, 5, 7, 11, 13, 17, 19\}$

$$Probability = \frac{number\ of\ desired\ outcomes}{number\ of\ total\ outcomes} = \frac{8}{20} = \frac{2}{5}$$

24) Choice C is correct

$615 \div 4 = \frac{615}{4} = \frac{600+10+5}{4} = \frac{600}{4} + \frac{10}{4} + \frac{5}{4}$

25) Choice C is correct

Perimeter of figure A is: $2\pi r = 2\pi\frac{20}{2} = 20\pi = 20 \times 3 = 60$

Area of figure B is: $5 \times 12 = 60$, $Average = \frac{60+6}{2} = \frac{120}{2} = 60$

SSAT Upper Level Mathematics Practice Test 2 Section 1

1) Choice C is correct

Digit 9 is in the hundredths place.

2) Choice A is correct

$x - 20 = -20 \rightarrow x = -20 + 20 \rightarrow x = 0$, Then; $x \times 20 = 0 \times 20 = 0$

3) Choice C is correct

$0.04 \times 13.00 = \frac{4}{100} \times \frac{13}{1} = \frac{52}{100} = 0.52$

4) Choice E is correct

His average speed was: $\frac{3.5}{0.5} = 7$ miles per hour

5) Choice B is correct

The perimeter of the quadrilateral is: $7 + 21 + 10 + 32 = 70$

6) Choice D is correct

The amount of flour is: $M - 75$

7) Choice D is correct

Number of packs needed equals to: $\frac{18}{4} \cong 4.5$, Then Mia must purchase 5 packs.

8) Choice C is correct

The time it takes to drive from city A to city B is: $\frac{3{,}600}{70} = 51.42$, It's approximately 51 hours.

9) Choice B is correct

Number of males in the classroom is: $60 - 30 = 30$

Then, the percentage of males in the classroom is: $\frac{30}{60} \times 100 = 0.5 \times 100 = 50\%$

10) Choice B is correct

For one person the total cost is: \$9.20, Therefore, for five persons, the total cost is:

$$5 \times \$9.20 = \$46$$

11) Choice B is correct

Let x be the fourth quarter rate, then: $\frac{93+40+87+x}{4} = 97$, Multiply both sides of the above equation by 4. Then: $4 \times \left(\frac{93+40+87+x}{4}\right) = 4 \times 97 \rightarrow 93 + 40 + 87 + x = 388 \rightarrow 220 + x = 388$, $\rightarrow x = 388 - 220 = 168$

12) Choice B is correct

Let x be the number. Write the equation and solve for x. $\frac{2}{3} \times 15 = 10 \rightarrow 10 = \frac{2}{5}x$, multiply both sides of the equation by $\frac{5}{2}$, then: $10 \times \frac{5}{2} = \frac{2}{5}x \times \frac{5}{2} \rightarrow x = 25$

13) Choice A is correct

Use the information provided in the question to draw the shape.

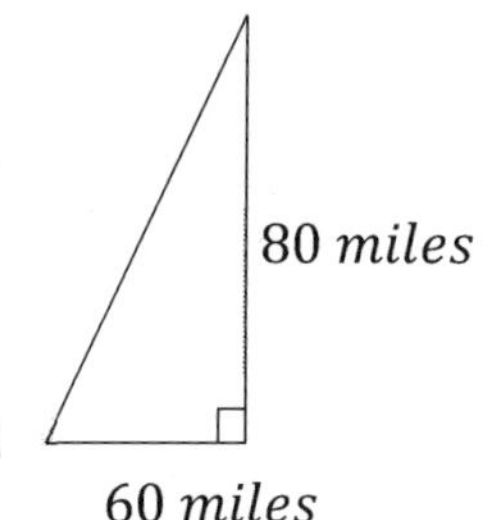

Use Pythagorean Theorem: $a^2 + b^2 = c^2$

$60^2 + 80^2 = c^2 \Rightarrow 3{,}600 + 6{,}400 = c^2 \Rightarrow 10{,}000 = c^2 \Rightarrow c = 100$ *miles*

14) Choice C is correct

Three times of 25,000 is 75,000. One sixth of them cancelled their tickets.

One sixth of 75,000 equals 12,500 ($\frac{1}{6} \times 75{,}000 = 12{,}500$).

62,500 ($75{,}000 - 12{,}500 = 62{,}500$) fans are attending this week

15) Choice A is correct

The equation of a line in slope intercept form is: $y = mx + b$, Solve for y: $8x + y = 14 \Rightarrow y = -8x + 14$,The slope of this line is -8.The slope of the line perpendicular to this line is:

$m_1 \times m_2 = -1 \Rightarrow -8 \times m_2 = -1 \Rightarrow m_2 = \frac{1}{8}$

16) Choice A is correct

Let x be the number of years. Therefore, \$3,000 per year equals $3{,}000x$. Starting from \$34,000 annual salary means you should add that amount to $3{,}000x$. Income more than that is: $I > 3{,}000x + 34{,}000$

17) Choice B is correct

The length of MN is equal to: $4x + 6x = 10x$, Then: $10x = 50 \rightarrow x = \frac{50}{10} = 5$

The length of ON is equal to: $6x = 6 \times 5 = 30\ cm$

18) Choice C is correct

Let the number of women should be added to city D be x, then:

$\frac{528 + x}{550} = 1.2 \rightarrow 528 + x = 550 \times 1.2 = 660 \rightarrow x = 132$

19) Choice B is correct

Ratio of women to men in city A: $\frac{570}{600} = 0.95$

Ratio of women to men in city B: $\frac{291}{300} = 0.97$

Ratio of women to men in city C: $\frac{665}{700} = 0.95$

Ratio of women to men in city D: $\frac{528}{550} = 0.96$

0.97 is the maximum ratio of woman to man in the four cities.

20) Choice D is correct

Percentage of men in city $= \frac{600}{1,170} \times 100 = 51.28\%$, Percentage of women in city $C = \frac{665}{1,365} \times 100 = 48.72\%$, Percentage of men in city A to percentage of women in city $C = \frac{51.28}{48.72} = 1.05$

21) Choice D is correct

Smallest 5–digit number is 10,000, and biggest 5–digit number is 99,999. The difference is: 89,999

22) Choice B is correct

$\frac{x}{3+4} = \frac{y}{11-8} \rightarrow \frac{x}{7} = \frac{y}{3} \rightarrow 7y = 3x \rightarrow y = \frac{3}{7}x$

23) Choice A is correct

$\sqrt[5]{x^{21}} = \sqrt[5]{x^{20} \times x} = \sqrt[5]{x^{20}} \times \sqrt[5]{x} = x^{\frac{20}{5}} \times \sqrt[5]{x} = x^4 \sqrt[5]{x}$

24) Choice C is correct

Perimeter of rectangle A is equal to: $2 \times (10 + 6) = 2 \times 16 = 32$

Perimeter of rectangle B is equal to: $2 \times (6 + 4) = 2 \times 10 = 20$

Therefore: $\frac{20}{32} \times 100 = 0.625 \times 100 = 62.5\% \cong 62\%$

25) Choice E is correct

The average speed of John is: $160 \div 4 = 40\ km$, The average speed of Alice is: $190 \div 5 = 38\ km$, Write the ratio and simplify. $40 : 38 \Rightarrow 20 : 19$

SSAT Upper Level Mathematics Practice Test 2 Section 2

1) Choice C is correct

$0.44 \times 12.8 = \frac{44}{100} \times \frac{128}{10} = \frac{44 \times 128}{100 \times 10} = \frac{5,632}{1,000} = 5.632$

2) Choice B is correct

$4\frac{3}{8} \times 5\frac{1}{9} = \frac{35}{8} \times \frac{46}{9} = \frac{35 \times 46}{8 \times 9} = \frac{1,610}{72} = \frac{805}{36} = 22\frac{13}{36}$

3) Choice E is correct

A. $\frac{3}{2} \times \frac{5}{9} = \frac{5}{6}$ is not a whole number

B. $\frac{1}{3} + \frac{1}{4} = \frac{4+3}{12} = \frac{7}{12}$ is not a whole number

C. $\frac{40}{6} = \frac{20}{3} = 6.66$ is not a whole number

D. $3.5 + 2 = 5.5$ is not a whole number

$3.5 + \frac{4}{8} = 3.5 + 0.5 = 4$ is a whole number

4) Choice D is correct

8 cubed is: $8 \times 8 \times 8 = 64 \times 8 = 512$

5) Choice A is correct

Let x be the number, then; $\frac{2}{3}x = 16 \rightarrow x = \frac{3\times16}{2} = 24$, Therefore: $\frac{5}{3}x = \frac{5}{3} \times 24 = 40$

6) Choice D is correct

Use formula of rectangle prism volume. $V = (length)(width)(height) \Rightarrow 3{,}000 = (30)(10)(height). \Rightarrow height = 3{,}000 \div 300 = 10$ feet

7) Choice D is correct

Number of boxes equal to: $\frac{450}{25} = \frac{90}{5} = 18$

8) Choice D is correct

The angle x and 60 are complementary angles. Therefore: $x + 60 = 180$, $180° - 60° = 120°$

9) Choice E is correct

Use simple interest formula: $I = prt$ (I = interest, p = principal, r = rate, t = time)

$I = (13{,}500)(0.045)(2) = \$1{,}215$

10) Choice C is correct

$\frac{504}{600} = 0.84 = 84\%$. 504 is 84% of 600. Therefore, the discount is: $100\% - 84\% = 16\%$

11) Choice D is correct

Use distance formula: $Distance = Rate \times time \Rightarrow 527 = 62 \times T$, divide both sides by 62.

$\frac{527}{62} = T \Rightarrow T = 8.5\ hours$. Change hours to minutes for the decimal part.

$0.5\ hours = 0.5 \times 60 = 30\ minutes$

12) Choice C is correct

The rate of construction company $= \frac{40\ cm}{1\ min} = 40\frac{cm}{min}$

Height of the wall after $50\ min = \frac{40\ cm}{1\ min} \times 50\ min = 2{,}000cm$

Let x be the height of wall, then $\frac{2}{3}x = 2{,}000\ cm \rightarrow x = \frac{3\times2{,}000}{2} \rightarrow x = 3{,}000\ cm = 30m$

13) Choice C is correct

Let's review the choices provided:

A. $28 - 2 = 26 \rightarrow \frac{26}{2} = 13 \neq 3$

B. $28 - 5 = 23 \rightarrow \frac{23}{5} = 4.6 \neq 3$

C. $28 - 7 = 21 \rightarrow \frac{21}{7} = 3 = 3$

D. $28 - 12 = 16 \rightarrow \frac{16}{12} = 1.33 \neq 3$

E. $28 - 28 = 0 \rightarrow \frac{0}{28} = 0 \neq 3$

14) Choice C is correct

Speed of car A is: $\frac{700}{7} = 100$ miles per hour , Speed of car B is: $\frac{700}{5} = 140$ miles per hour

$\rightarrow 140 - 100 = 40$ miles per hour

15) Choice C is correct

x is the number of all sales profit and 4% of it is: $4\% \times x = 0.04x$, Employee's revenue: $0.04x + 8000$

16) Choice B is correct

$\frac{1}{5} \cong 0.2$ $\frac{5}{3} \cong 1.66$ $\frac{8}{11} \cong 0.73$ $\frac{2}{3} = 0.66$

$$\frac{5}{3} > \frac{8}{11} > \frac{2}{3} > \frac{1}{5}$$

17) Choice D is correct

The ratio of boys to girls is $3:2$. Therefore, there are 3 boys out of 5 students. To find the answer, first divide the total number of students by 5, then multiply the result by 3.

$700 \div 5 = 140 \Rightarrow 140 \times 3 = 420$

18) Choice E is correct

$\frac{(8+6)^2}{2} + 6 = \frac{(14)^2}{2} + 6 = \frac{196}{2} + 6 = 98 + 6 = 104$

19) Choice D is correct

$y = 5ab + 2b^4$, Plug in the values of a and b in the equation: $a = 3$ and $b = 2$

$y = 5(3)(2) + 2(2)^4 = 30 + 2(16) = 30 + 32 = 62$

20) Choice B is correct

Use the formula of the area of circles. $Area = \pi r^2 \Rightarrow 49\pi = \pi r^2 \Rightarrow 49 = r^2 \Rightarrow r = 7$

Radius of the circle is 7. Now, use the circumference formula: Circumference $= 2\pi r = 2\pi(7) = 14\pi$

21) Choice D is correct

$0.8x = (0.4) \times 20 \rightarrow x = 10 \rightarrow (x+7)^2 = (10+7)^2 = (17)^2 = 289$

22) Choice D is correct

Prime factorizing of $18 = 2 \times 3 \times 3$, Prime factorizing of $24 = 2 \times 2 \times 2 \times 3$

$GCF = 2 \times 3 = 6$

23) Choice A is correct

The sum of all angles in a quadrilateral is 360 degrees. Let x be the smallest angle in the quadrilateral. Then the angles are: $x, 2x, 3x, 4x,$

$x + 2x + 3x + 4x = 360 \rightarrow 10x = 360 \rightarrow x = 36$, The angles in the quadrilateral are: $36°, 72°, 108°$, and $144°$, The smallest angle is 36 degrees.

24) Choice E is correct

Length of the rectangle is: $\frac{4}{5} \times 30 = 24$, Perimeter of rectangle is: $2 \times (24 + 30) = 108$

25) Choice A is correct

$\frac{3}{4}$ of $160 = \frac{3}{4} \times 160 = 120$, $\frac{2}{3}$ of $120 = \frac{2}{3} \times 120 = 80$

"Effortless Math Education" Publications

Effortless Math authors' team strives to prepare and publish the best quality SSAT Upper Level Mathematics learning resources to make learning Math easier for all. We hope that our publications help you learn Math in an effective way and prepare for the SSAT Upper Level test.

We all in Effortless Math wish you good luck and successful studies!

Effortless Math Authors

www.EffortlessMath.com

... So Much More Online!

- ✓ FREE Math lessons
- ✓ More Math learning books!
- ✓ Mathematics Worksheets
- ✓ Online Math Tutors

Made in the USA
Columbia, SC
01 July 2022